9-1 -1 4

1·30 - 5·00 4½ R.

6·30 - 10·30 4

9 - 12·00. Wyn.

12 - 3·30. 4eth.

3·30 - 6·00 Eng.

6·30 - 8·00 gocs

8·00 11·00. gocs.

*Unwin Education Books: 29*

THE FOUNDATIONS OF MATHS IN THE INFANT SCHOOL

*Unwin Education Books*

Series Editor: Ivor Morrish, BD, BA, Dip.Ed. (London), BA (Bristol)

1 Education Since 1800 IVOR MORRISH
2 Moral Development WILLIAM KAY
3 Physical Education for Teaching BARBARA CHURCHER
4 The Background of Immigrant Children IVOR MORRISH
5 Organising and Integrating the Infant Day JOY TAYLOR
6 The Philosophy of Education: An Introduction HARRY SCHOFIELD
7 Assessment and Testing: An Introduction HARRY SCHOFIELD
8 Education: Its Nature and Purpose M. V. C. JEFFREYS
9 Learning in the Primary School KENNETH HASLAM
10 Sociology of Education: An Introduction IVOR MORRISH
11 Fifty Years of Freedom RAY HEMMINGS
12 Developing a Curriculum AUDREY and HOWARD NICHOLLS
13 Teacher Education and Cultural Change H. DUDLEY PLUNKETT and JAMES LYNCH
14 Reading and Writing in the First School JOY TAYLOR
15 Approaches to Drama DAVID A. MALE
16 Aspects of Learning BRIAN O'CONNELL
17 Focus on Meaning JOAN TOUGH
18 Moral Education WILLIAM KAY
19 Concepts in Primary Education JOHN E. SADLER
20 Moral Philosophy for Education ROBIN BARROW
21 Beyond Control? PAUL FRANCIS
22 Principles of Classroom Learning and Perception RICHARD J. MUELLER
23 Education and the Community ERIC MIDWINTER
24 Creative Teaching AUDREY and HOWARD NICHOLLS
25 The Preachers of Culture MARGARET MATHIESON
26 Mental Handicap: An Introduction DAVID EDEN
27 Aspects of Educational Change IVOR MORRISH
28 Beyond Initial Reading JOHN POTTS
29 The Foundations of Maths in the Infant School JOY TAYLOR

*Unwin Education Books: 29*
*Series Editor*: Ivor Morrish

# The Foundations of Maths in the Infant School

JOY TAYLOR
*Head of the Department of Education*
*Oxford Polytechnic*

London George Allen & Unwin Ltd
Ruskin House Museum Street

First published in 1976



ISBN 0 04 372014 5 hardback
0 04 372015 3 paperback

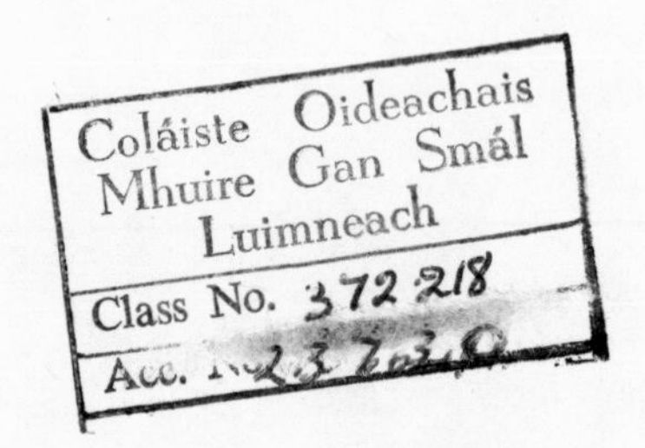


Set in 10pt Times Roman type
Composition by Linocomp Ltd
Marcham, Oxon
Printed in Great Britain by
Biddles Ltd, Guildford, Surrey

*To*
*Micky and Andy*

*Acknowledgements*

I warmly acknowledge a debt of gratitude to Les Holland, whose help on the final revision of the text has made this a better book than it would otherwise have been; to Dorothy Evans, Derrick Field and Frank Carter, for their generous advice on certain aspects of mathematics and psychology; and to Joan Webb and her colleagues in the Lady Spencer-Churchill College library, who have never failed to turn up the books and other publications I have needed.

## *Preface*

This is not, primarily, a book about mathematics. It is about children learning mathematics; about the steps the teacher may take to help them with their learning, and why she might consider taking this step rather than that. It therefore sets out to analyse, in some detail, the nature and progression of mathematics as they apply to young children, and to relate them to everyday classroom practice. The purpose of the book is to try to provide a bridge between the theoretical statement and the source of ideas about mathematical activities, while acknowledging that both are necessary for the successful teaching of the subject.

The range of ability in an infant school can be so wide that any book which attempted to cover as extensive a subject as mathematics throughout the full ability range would need to be a good deal longer than this one. It therefore stops short of the kind of maths in which the really advanced child might engage towards the top of the infant school. There is already more literature available for the later stages than for the earlier ones. Many children, of course, will go no further in the infant school than the stages described in these pages, and indeed some will not go as far. The concentration, however, is on the foundation skills of mathematics, and on the first steps that may be taken towards building upon them.

For convenience, the words 'class' and 'classroom' are used throughout to denote the group of children for whose learning the teacher is responsible, and the school space in which she and the children operate. The 'classrooms' may or may not be physically separated from each other, and the teacher may be wholly responsible for the work of her 'class' or she may be a member of a team sharing the teaching of a larger number of children.

The practice is presented in the order of expected levels of development, while recognising that with a wide ability range – which is even wider if vertical grouping is in operation – only some children in the class will be at the level described at any one time. Certain aspects of practice and organisation, however, continue to have application beyond the level in which they are described. In order to avoid tedious repetition, this continuation is assumed rather than detailed all over again.

These devices are used only for the sake of clarity and brevity. The teacher will appreciate that, whichever way material is presented in a book, some adaptation is always necessary in order to apply it to a particular teaching situation.

Extensive references for further reading are given in the text, which seeks to do no more than suggest a practical line of advance, supported by the essential theoretical background. The teacher's continuing study and her increasing experience are the two critical factors which contribute more than anything else to the children's steady progress in mathematics.

J.M.T.

*Contents*

PART ONE

# The Theoretical Arguments

*Chapter 1*

# The Old and the New

Along with the teaching of reading, the teaching of mathematics in the primary school has probably received more publicity in the last ten or twenty years than has ever before been given to the curriculum in the infant or junior school. This has been due partly to the widespread expression of the view that mathematics as it used to be taught failed to give enough children a sufficiently sound understanding of the subject to meet the needs of an increasingly technological age; and partly to the fact that the 'new maths' quickly reached parents through homework, and the general public was much aware of the immense change in the teaching of mathematics that was taking place.

The publicity drew attention to the considerable effort that was being made to dispense with the meaningless conundrums familiar to most parents when they were at school. Even in 'those days' it was very difficult to be interested in how long it took to fill the bath when you failed to put in the plug. On the other hand, the publicity also aired the views of the 'when I was at school I was made to get down to it and do some work' brigade, many of whom were genuinely concerned that their children did not appear to know their tables despite the 'fancy' graphs and Venn diagrams that came home from school in the evenings.

As with any new method being tried on a large scale in school, the extremes of theory and practice were (and still are) to be found, together with everything in between. At one end of the spectrum there are the enthusiasts who decorate the walls with sets and graphs and who feel positively ill if they inadvertently come upon a 'plus' or an 'equals' sign; and as for tables, well – they went out with the ark and we really don't do that sort of thing here. At the other extreme there are still the dedicated chalk and talk merchants, who believe that a proper grounding in maths can be achieved only if every child, at a very early stage in his school life, doggedly chews his way through every sum card in the wall pockets which march round the room from the door to the far end of the blackboard.

But somewhere between these not wholly fanciful outposts of the mathematical empire, solid progress is being made in helping children really to participate in the mathematics of their daily lives and to apply fundamental mathematical concepts with interest and understanding. Some children, indeed, can go very much further than this, and can set out along the path of learning to use mathematics as the sophisticated discipline that it is. This comes later than in the infant school, but it is there where lifelong attitudes are often established and where failure to understand may set up a barrier through which a child never really passes. It is therefore of crucial importance that the infant teacher should be clear about what she is trying to do and why she is trying to do it. She must know what her objectives are, or she cannot lay a solid foundation on which the future edifice – however large or small – can be built.

It is perhaps worth considering at this point precisely what mathematical skills every adult needs for the satisfactory conduct of his or her daily life, because it is basically these skills which many children first meet in the infant school. This is not to say that all the mathematics an adult needs can be learned by all children in their first two or three years in school. Certainly this early learning has to be consolidated and extended, and it is true that to some the acquisition of even the most elementary skills takes a long time. However, the basic mathematical concepts which many children encounter in their early school years are those which ultimately provide the skeleton of mathematics that is essential to every adult if he is not to be severely handicapped.

This essential skeleton should enable men and women to budget accurately enough to make ends meet; to calculate the cost of purchases when they go shopping and to know whether their change is correct; to estimate the quantity that will be available for each person when dividing food or other commodities among members of the family; to measure the length of material needed to curtain the windows or to provide wood for shelves and to arrive at a calculation of the cost of so many metres at a given price; to understand the unit of measurement in which potatoes or milk or petrol are bought and to estimate the quantity represented by a number of such units; and to tell the time and estimate the duration of time as they go about their daily tasks. Prerequisites of all this are an understanding of the counting system and a knowledge of the vocabulary of simple mathematics.

It is true, of course, that some adults are not able to do all these things, and they survive; but this does not mean that we should not try to equip everybody with at least these basic tools. It is also true that, to a limited extent, some of this mathematical activity can be undertaken for people by others. The girl in the supermarket will add the cost of your purchases on her till and give you the change. But there are times when there is no girl and no till and you are forced to make money calculations for yourself. It helps if you can do this fairly easily and quickly, and it would undoubtedly be a most damaging disadvantage if you could not do it at all. There is, of course, a great deal more to mathematics than the elementary skills just enumerated and many people make good use of maths of a much more complex nature. There are, too, many elements of mathematics which are not utilitarian and which in themselves give enjoyment and artistic satisfaction. However, the point at issue is that the minimal amount of mathematics that adults really cannot do without is, in essence, that which is first encountered in the infant school and which has to be learned and taught despite the changes that have come about in modern life. It is simply not true to say that we don't need to be able to compute any more, because there are machines to do it for us; or that learning tables is a useless waste of time because we never need them in adult life. Certainly we need them. Can one imagine having to use counters or make notches in sticks in order to arrive at the number of tiles needed to do the wall above the bath, having found that there will have to be four rows with twelve in each row? And how many people have calculating machines available for the kind of everyday computation which most of us find ourselves making from time to time?

It is hardly likely, of course, that anyone would argue against the necessity for equipping all children with at least the basic mathematical skills. But there are indeed those who believe that with the advent of 'modern maths' and 'progressive' teaching and organisation, children will experience such stimulating learning situations that few will fail to absorb all the mathematics they need, and more. Moreover, they will learn to 'think mathematically' and be able to apply this thinking to the rapidly changing demands of the modern world. So sums and tables are out, and operational games and projects are in, and we learn mathematics by counting the number of beetles we find in two square metres of grass when doing environmental studies.

These views are given weight by the undeniable truth that so much of the 'old maths' failed to do the job anyway. The HCFs and LCMs, the hours spent on long division of money, the rote learning of eight sevens are fifty-six when half the children did not understand the mathematics of the exercise – this sort of activity was wasteful and unproductive and did not help anyone to make use of mathematics, let alone enjoy it. But the unhappy fact is that all the publicity about the new and exciting ways of learning maths, combined with the pressures to abandon outdated and rigid subject barriers, have left many teachers of young children unnecessarily bewildered about what they should or should not do. Is it wrong to get Fred to weigh six cotton reels after all? Or to do sums? Or to buy a pre-arranged list of items from the classroom shop and add up the cost? Or to learn how to multiply?

One of the consequences of this uncertainty has been that teachers, particularly in infant schools, have flocked to in-service courses on the teaching of mathematics in order to learn how it should be done; and they are sometimes disappointed because what they are told is either impractical with a class of thirty-five children, or else it is not so very different from what they are already doing anyway. This is not an indictment of the many excellent in-service courses on the teaching of mathematics that are provided in many places, which successfully fulfil the purpose of refreshing and extending the teacher and giving her new ideas that she can beneficially use with her own class. But it is an indication that a very large number of infant teachers have the impression that something revolutionary has happened in the world of mathematics, and if they do not join the revolutionary forces the children they teach are forever doomed.

There is, however, a more realistic view of the situation. The revolution has certainly taken place. But its effects are much less dramatic in infant schools than they are where older children are concerned. For a great many years, the majority of infant teachers have been using individual and practical methods for the teaching of mathematics, and the children have actively engaged in situations from which they have learned. Certainly there is still some formal teaching of the unproductive kind, and schools can be found where children are engaging in activities and exercises without the necessary foundation of mathematical understanding. Nevertheless, the methods that have long been practised in a great many infant schools need only evolutionary development to keep up with the 'revolution' that has extended up the age range. The changes which have made the most impact in the early school years are in two main areas:

1. In methods of presenting and recording mathematics, e.g. the use of arrow diagrams, block graphs and other forms of graphical representation.
2. In the introduction of additional mathematical materials and situations from which young children can profitably learn, e.g. logic blocks and other structural apparatus, and the more extensive use of carefully devised mathematical games.

In addition, of course, more is always being learned about how and when children are thought to acquire mathematical concepts and this throws light on the methods we should use in helping them to do so. However, although it would be capricious to under-rate the value of the innovation in infant schools, the 'revolution' that has taken place is more in the nature of the extension and sometimes the modification of established methods, rather than of wholesale change.

One of the important characteristics of mathematics is its sequential nature and it is imperative that the teacher should be aware of this. Certain concepts must be grasped before others can be approached and material should be presented in a rational progression. It is for this reason that mathematics cannot be learned entirely through projects. The mathematical possibilities of a project should indeed be explored and use should be made of the opportunities for learning that arise naturally. But it cannot be supposed that such opportunities will entirely serve the needs of the sequential learning that is necessary, and mathematics must be presented systematically if the child's understanding is to be sound. Nothing in 'modern' maths alters a requirement that is inherent in the subject itself.

The pattern of progression adopted in this book is to divide it into two main parts:

*1. The Beginning of Numeracy.* Here we are concerned at first with the child for whom numbers have little or no quantitative significance, though he uses some of the vocabulary of mathematics and has a foundation of mathematical experience. Many children have progressed beyond this stage when they begin school, but others have not and the teacher must make suitable provision for them.

We continue through the stage when counting to ten or thereabouts becomes secure, with an understanding of one-to-one correspondence. Number values to about ten come to have quantitative meaning and these numbers are recognised when spoken and written. Conservation of number may at first be grasped spasmodically, and is finally securely established.

*2. Building on the Foundation Skills.* Numbers to ten or twelve can now be grouped together and split up, but concrete material such as rods or counters of some kind are necessary.[1] Notation to ten or twelve is established and the counting system can be extended to twenty and, in a limited context, beyond. Within fairly close limits the child is now ready to undertake number operations.

We then move on to the stage when more varied operations, and with larger numbers, are possible. Children at this level should be reading pretty fluently, so number operations should also be presented to them in words. Counters may be discarded by those who no longer find them necessary because they are capable of some degree of abstract thought. By no means all children will reach this stage in the infant school. On the other hand, there may be some whose mathematical understanding is sufficiently advanced for them to progress a good deal further.

Although these levels of progression have just been described in terms of working with numbers, the children will naturally be engaging in many other aspects of mathematics at the same time. These aspects are included in the text at each level but, for the sake of clarity, they are given under separate headings of length (and, where appropriate, two- or three-dimensional measurement), substance or mass and liquid measure, weight, time, money, and other mathematical activity. This is not to suggest that mathematics should in practice be separated into compartments which do not connect with each other. Clearly they do connect, but for clarity of presentation they are dealt with in this way, and the kind of activity suitable for children at each of the levels of development is indicated.

It is perhaps stating the obvious to say that if children are to make systematic progress in mathematics, to understand what they are doing and to enjoy doing it, they must be given constructive and informed teaching. But what *is* 'teaching' in this context? We know that it is no longer a matter of imparting a pre-determined body of knowledge to a whole class of children, and then setting endless exercises in the belief that practice makes perfect. And it is emphatically not just a question of presiding over a stimulating environment while the children discover it all for themselves. We need a definitive statement which falls between these two extremes and which gives the teacher a positive indication of her function when teaching mathematics to children of infant school age.

A practical statement of her function is that she should:

1. Know as much as possible about the kind of mathematical concepts the young child is thought to be able to acquire.
2. Recognise the stage at which he is likely to acquire them or be helped to acquire them.
3. Make available to him, at the right time and in the right order, the kind of activity and materials and experience that will enable him to learn what she is seeking to teach him.
4. Ensure that she gives him direct help and teaching when he needs it and the opportunity to consolidate his learning with practice when this is appropriate.

These four requirements, which are of theoretical as well as practical significance, cover an extensive field, and they make heavy demands upon a teacher's professional skill. But they are of critical importance in laying the foundation of numeracy with young children. Moreover, unless the sequential nature of mathematics is recognised there will be cracks in the foundation and the future structure will be weakened. This does not mean that learning maths must be a wearisome labour. To present material systematically does not limit the nature of the material to that which is burdensome. The imaginative teacher will not impoverish the children's learning, but will make good use of the rich resources that mathematics can provide.

## NOTE

1 The word 'counters' is used thoroughout in a general sense, and does not imply that counters must necessarily be of the conventional plastic or card-board variety. Indeed, such things as acorns, pegs, bottle tops, etc. are often better, because they are easier for a small child to pick up from a flat surface.

*Chapter 2*

# Introductory Considerations



Few infant teachers would claim to be specialist mathematicians in the conventional sense of the word, but they nevertheless recognise that they cannot effectively teach mathematics to young children unless they have an understanding of its principles and an appreciation of their implications in the classroom. This understanding cannot be derived from reading any one book, however comprehensive, and in any case there is a place for detailed academic studies based on practical research and also for the interpretation of such studies in day-to-day classroom practice. But advice on classroom practice, if it is no more than a list of things to do, does not help the teacher to know why she is doing the things on the list. And if she does not know why she is doing them, she has no means of selecting from the list those items which are appropriate to a particular child who has reached a particular level of understanding.

So it is necessary to introduce the practice against a theoretical background. Further study adds substance to the background, and experience enriches the practice. The purpose of this chapter is to outline some of the theoretical arguments in so far as they underpin the practice in the infant school, and to draw attention to the more detailed study in which the teacher should engage if she is to teach mathematics to young children to real effect.

To many teachers who have trained in recent years, there appears to be one especial god who presides over the world of mathematics, and this god is called Piaget. The Piagetian stages are imprinted on every young teacher's mind, and if she remembers nothing else from her mathematical training she remembers these. Certainly she ought to do so, for Piaget and his colleagues have had an immense impact, not only on the teaching of mathematics but on teaching generally. However, children are not well served if their teachers sit at the feet of only one god, and ignore all the others in the mathematical firmament. For there are other gods whose words are worth hearing, without diminishing the contribution that Piagetian theory has made. We will now engage in reflections upon several of the arguments with which the teacher should acquaint herself; but these will provide no more than a general introduction, or starting point for more intensive study.

What then is known, or suggested, about the formation of early

mathematical concepts in children? It is true to say that much has yet to be learned about how, and when, young children form concepts, including those which are specifically applicable to mathematics. We do not really know, for instance, that all children arrive at the same concept in the same way. It seems that some children may need one kind of experience, and others may need something else. Indeed it is very probable that most children need a variety of experience, and the teacher would be wise to provide this variety if children of widely differing temperaments and capacities are to learn mathematics and enjoy doing so.

It may be, for example, that for one child a strongly motivational approach is the most successful. That is to say, he needs an approach based on actual or simulated real-life situations – giving out cups of tea in the Wendy House (one for each person, with corresponding saucers and spoons), using the classroom shop, counting the milk bottles or the boys wearing jeans. These are activities in which the motivation is provided by the interest of familiar situations in and around the classroom, and by the child's personal involvement in them.

For another child, however, the motivation may come largely from structural materials which he handles and compares, and with which he can, in a sense, actually 'see' the mathematics happening. In this way he is ultimately able to conceptualise the basic mathematical processes. And again, there may be certain structured situations, by means of which the child learns through 'playing games' that enable him to extract the essential foundations of his mathematical understanding.

Apart from the fact that most children probably benefit from a variety of approach anyway, the teacher will also wish to make use of different kinds of materials and activities in accordance with the nature of the mathematics she wants the child to learn. So here at once we arrive at a situation in which she has to make choices. Upon what hypotheses are these choices to be based?

The first thing we must realise is that there is more than one view about how children form concepts. Most teachers are hardly in a position to determine that one view is correct and another is not, so they must acquaint themselves as far as possible with the theory and try to allow for its variations to be admitted in practice. Lovell gives his own clear analysis of concept formation in his invaluable book *The Growth of Basic Mathematical and Scientific Concepts in Children.*[1] He explains that the traditional view of concept formation is that the child begins with the *percept*, or interpretation by the brain of stimuli received through the senses from the external world; and from a very early stage he begins to abstract from his perceptions certain generalisations concerning the information he receives from the environment. At first his abstraction of these generalisations is an unconscious mental process, but as he develops he is able to exercise his judgement and consciously abstract generalisation from his contact with the world around him. His perceptions have now become conceptualised: the *concept* is formed. The child's capacity to do this may be influenced by such things as parental interest and the quality of his experience, but:

'Abstraction and generalisation are essentially mental processes; they are carried out in the mind. Adults can arrange an environment which may help them forward but the child himself makes the jump from the percept to the concept.'[2]

Generalisation helps the child to categorise the information he receives from the external world. For example, a small child may meet with a four-legged animal which is described as a 'dog'. At that stage, *any* four-legged animal may, to the child, represent one category and elicit the same descriptive term. He may describe everything in this category as a dog. As he develops and his experience extends, he comes to discriminate between different kinds of four-legged animal. In time, it becomes clear to him that only certain of these four-legged animals possess those common properties which categorise them as dogs. He recognises that although there are differences between members of a class, some differences do not exclude them from that class but others do. For instance, dogs may be different in size or colour, but these differences do not exclude them from belonging to the class of dog. But they cannot purr or say 'miaow'. They cannot normally be ridden upon. They cannot be as big as a cow or as small as a mouse. These differences would exclude them from the class of dog.

In other words, the child has now arrived at the stage at which he is able to make an intellectual judgement based on the information he has abstracted from his environment. He has conceptualised a generalisation concerning the class of dog, and his language enables him to describe it as a class distinct from other classes of animal. Lovell defines a concept as:

'. . . a generalization about data which are related. . . . Concepts seem to arise out of perceptions, out of actual acquaintance with objects and situations, and through undergoing experiences and engaging in actions of various kinds.'[3]

He points out that concept formation is assisted by memory and imagery, by trial and error, and by language. And although there are indications that concept formation does not always proceed from the concrete to the abstract, there is no suggestion that young children do not need concrete experience in order to abstract the generalisations on which the formation of mathematical and other concepts depends.

We may consider for a moment the application of the view of concept formation just described to situations involving numbers. If we follow the analogy of the conceptualisation of the class of dog, we encounter no reason why concept formation in recognisably mathematical situations should not be subject to the same 'rules'.

The concept of number value is very specific. The class of 'two', for example, has only one common property – that of quantity. The two objects may be different in kind, in colour, in size, in fact in every respect other than that of quantity. There may be two objects on the table – a green apple and a white plate. If we are limiting the classification to things which are white, or are fruit or apples or plates, there is only one object which meets the requirement. But if we are looking specifically for a quantitative property, that property in this case is two; and the same would apply to any two objects that we wished to classify in that manner.

Now it seems that the child's experience and actions will lead him from the state of not making any particular quantitative association with the objects (except, perhaps, a very vague one such as 'many'), to the state of recognising that individual objects may be assembled in collections of variable quantity, and that only *one* of these quantities can be described as 'two'; moreover, that this particular quantity is invariably described as 'two'.

To grasp this concept of number value seems to be far more difficult for some children than it was thought to be at one time; and the methods that the teacher uses, and the kind of teaching that she gives the child, may help him to grasp this concept to a greater or lesser extent. According to Lovell:

'Unfortunately little is known about the exact ways in which concept formation can be speeded up although we know that environmental conditions are of importance. Lovell (1955) showed that adolescents and young adults from a stimulating background were, in ability to categorize and form fresh concepts, superior to those from less favourable backgrounds – this after making allowance for general intelligence or academic aptitude. Churchill (1958) showed that infant children who had had opportunities to play with certain materials might develop certain mathematical concepts quicker than a control group who were not given such opportunities.'[4]

This seems to suggest an important theoretical proposition to which we find ourselves being drawn time and time again: *that the acquisition of concepts is not simply a question of maturation.* Other influences appear to contribute to it, and there is a strong case for arguing that one of these influences is teaching. For further reflections on this we turn now to the work of Dienes, whose studies give sound support to the substance of these propositions.

## NOTES

1 K. Lovell (London: ULP, third impression, 1971).
2 Ibid., p. 13.
3 Ibid., p. 13.
4 Ibid., p. 17.

*Chapter 3*

# Zoltan P. Dienes

If the acquisition of concepts is not simply a question of maturation, neither is it just a matter of providing children with an unstructured environment, rich and varied though it might be. Dienes attaches great importance to the influence of the environment, but it is one which is most carefully structured so that it induces the learning which is intended to take place.

It is perhaps appropriate at this stage to consider at least part of the theoretical base on which Dienes rests children's mathematical learning, though clearly the reader must refer to more specialised works for anything more than an outline of some of his arguments. As a psychologist, he has made a considerable study of the question of concept formation, and he suggests that personality factors influence not only concept formation itself but also the way in which children can make use of conceptualisation in further learning. He strongly advocates the use of *structured* play situations, in which the child moves gradually from a possibly intuitive grasp of the 'rules' (which Dienes does not despise, because he believes that much valuable learning begins this way), to a point at which the rules can be analysed and even manipulated to extend mathematical understanding. A challenging suggestion which emerges from Dienes's research is that children do not always learn more successfully if they are introduced first to a very simple mathematical structure and only later to a more complicated one.

Williams explains that this

'. . . seems to contradict many suppositions about the order in which material should be learnt. . . . It was found that, in some cases, more effective learning takes place when the learner is introduced first to a structure that is both more general and more complex, and only after this to a structure that is more particular and simpler.'[1]

The studies of Dienes and his colleagues led them to distinguish six stages in the process of learning mathematics.[2]

*First Stage*

'The notion of the environment seems to us to be of outstanding importance, for, in a certain sense, all learning is basically a process by which the organism adapts to its environment.'

This is the 'free play' stage, during which the child adapts to the structured environment that is provided for him. Before this adaptation takes place, he is badly adjusted to his environment. That is to say, he is at its mercy; he cannot exercise any control over the learning situations with which he is confronted. Ultimately, thanks to learning, he is able to dominate these situations.

'If we accept that it is the process of adaptation which represents all learning, it would seem reasonable to present the child with a suitable environment to which he might adapt himself if we wish learning to take place.'

In Dienes's 'suitable environment' the child is presented with carefully designed games and other learning situations, which will teach him certain things and enable him to form particular concepts. Though he is playing these games freely, he is doing a great deal more than simply amusing himself (though it is hoped that he is doing this as well).

'If we propose to teach logic to a child, it seems necessary to confront him with some situations which will lead him to form logical concepts.'

In considering logic:

'. . . we need to recognise that the natural environment in which the child lives does not embrace all the attributes which we consider as logical. It is necessary, then, to invent an artificial environment. As a consequence of this environment the child will be led little by little to form logical concepts in a more or less systematic way.'

So, although in this first stage the child is engaging in free play he is playing, not with anything that happens to come to hand, but with particular materials and with an aim in view that is pre-determined by the teacher. It may therefore be taken to follow that a structured programme, and organised and systematic learning, are necessary for the child if he is to form the mathematical concepts on which further progress is based.

*Second Stage*

In this stage the child comes to realise that the games impose constraints upon him. 'There are certain conditions which must be satisfied before certain goals can be attained.' So the child finds that there are 'rules of play'. These rules may be imposed by the teacher, or by the nature of the material or, in time, by the child himself; he then imposes his own constraints and adapts his game accordingly. Learning is taking place.

'Clearly, if one wishes the child to learn about mathematical structures, the sets of rules that one suggests will depend upon the relevant mathematical structures.'

Here again we see that the 'play' is designed with quite specific ends in view, and learning is not left to the chance contacts that may come a child's way in unstructured play situations.

*Third Stage*
Dienes does not suggest that children learn mathematics simply by playing games, however carefully devised.

'Obviously, to play structural games according to mathematical laws intrinsic in any mathematical structure is not *learning mathematics.* How is the child to be able to extract from this set of games the underlying mathematical abstractions?'

Dienes's answer to this question is that children, by playing different games with similar underlying structures, come to isolate the common element (or 'underlying mathematical abstraction'); and they then recognise the abstract links in games that *appear* to be different from each other.

'Thus, the games played with one embodiment, then again with another embodiment, will be identified from the point of view of their structures. It is at this moment that the child realizes what is the "same" in the different games that he has experienced, that he will have made an "abstraction".'

The important question here is the implication that the teacher can assist the process of abstraction by providing the kind of material that is designed to lead the child towards a state of understanding. Thus it is not merely a matter of making varied provision and then waiting for the moment of abstraction to arrive. The teacher, it seems, may be able to help it to arrive.

*Fourth Stage*
The child cannot yet make use of this abstraction, 'because it is not yet properly fixed in his mind'. For this to happen, he needs to be able to represent the abstraction in some form. This may be a visual representation, such as a graph or a diagram, but for children who do not think

'in an essentially visual way' an auditory representation may be more suitable. The point at issue in this fourth stage is that until the child has represented his abstraction, he cannot examine it and talk about it and see how he might apply it in another situation. This would seem to add an important dimension to the question of recording generally. If we accept Dienes's view, it follows that 'doing', by itself, is not enough. The learning from the doing needs to be *recorded* if the child is to derive full value from it.

*Fifth Stage*

The child can now examine his representation, or even several different representations of the same abstraction. 'The goal of this examination is to understand some properties of the abstraction which has been made.' To achieve this goal the child must be able to describe what he finds in the course of the examination; and description requires a language. So, in this fifth stage, a language must be invented to describe the representation (fourth stage) of the properties of the abstraction which emerged in the third stage. Dienes suggests that it is valuable for children to invent their own descriptive language which reflects their own ideas; and then, in discussion, children and teacher can decide upon the language most advantageous for a particular description. 'Such a description forms the basis for a system of axioms' – or accepted principles – which are then available for use as learning progresses.

*Sixth Stage*

This is the stage in which language and mathematical activity make further advances, and 'a formal system' is devised. Since there are so many properties in mathematical structures, it is not possible to include them all in one description. So one takes as few descriptions as possible, and one uses them as starting points from which other properties may be deduced.

'The minimum number of these descriptions are called the *axioms*. The process of deducing others is called *proof* and these latter properties are called *theorems*.

The manipulation of such a system, called a formal system, is the final goal of the mathematical study of a structure.'

In other words, one-to-one correspondence, when it is grasped, may provide the child with an *axiom*; the recognition that two is 'one more than one' is a *proof*; and the numbers which he can then construct are the *theorems*. He is now equipped with the beginnings of *a formal system* which enables him to take further steps in learning mathematics.

A teacher may feel that the development of Dienes's theory concerning the 'six stages' has not yet reached a point at which enough of it has been subjected to sufficient 'clinical testing', in ordinary classroom conditions, to give it immediate day-to-day application in her teaching of mathematics. This is not to deny the validity of the work of a man of Dienes's stature. It is simply to recognise that there usually is, for a while, a natural gap between what can be done, in controlled conditions, by a leading psychologist and mathematician with an international reputation and presumably a team of able and experienced research colleagues, and by an 'ordinary' teacher in an 'ordinary' school. However, having recognised that the gap may exist, we do well to remind ourselves that theoretical innovation which has ultimately revolutionised practice in very ordinary classrooms is far from unknown to teachers. Sooner or later the gap closes, and it happens sooner rather than later if the teacher, in a receptive frame of mind, really informs herself of the theory and then seeks to begin the process of application in terms which are realistic in her particular teaching situation.

It is for this reason that an attempt has been made to interpret Dienes's six stages in some detail; we cannot ignore the significance of his argument – which is that mathematics has traditionally been taught the wrong way round. For example, children may be introduced, at a fairly early stage, to symbols (such as numbers). In order to understand them, they need assistance from concrete materials of many kinds, and in trying to represent them they do not always grasp the abstract idea (e.g. the theorem of 'two') of what they are representing. So they are provided with games and other 'real life' situations to help them to understand what they are being asked to do. 'So, finally,' says Dienes, 'the child reaches the real life situation, which is where he should have started. Thus in traditional teaching, the direction of study is exactly the reverse of that which is proposed in these pages.'

It is perhaps pertinent at this point to consider what we understand by the term 'traditional teaching'. In its extreme form it presumably means introducing the young child to mathematical symbols and abstract ideas which he does not understand, in the belief that understanding will come as a result of practice. It is superfluous to state that teaching of this kind is not often found today in schools where children are at an early stage of learning mathematics. However, a number of questions do emerge which in all honesty we should ask ourselves, and to which Dienes's research findings seem particularly applicable.

Are we sure that we always give children enough experience of 'play' or 'experimental' activity, in appropriate environmental conditions (Dienes's first stage), before we ask them to systematise this activity by engaging in some form of symbolic representation? How structured is this play? Does it, in due course, reach the point when it enables the child to recognise that 'there are certain conditions which must be satisfied before certain goals can be obtained'? For example, when children are sorting things into sets, are the 'rules of play' now apparent, and the constraints recognised, so that the activity is mathematically productive (Dienes's second stage)? Do we then move to the point when mathematical activity is carefully enough designed and controlled for the child to be able to identify the same properties in different situations, and thus make an abstraction in regard to these properties (third stage)? Does the child then record in some way, in order to fix and examine the abstraction; then find the language necessary to describe it, and finally establish 'a formal system' in which the mathematical foundation is secure (fourth, fifth and sixth stages)?

It could be worth the teacher's while to re-examine her approach and see whether there are ways in which she believes that some modification or change of emphasis might have beneficial results in the children's learning. In practical terms, she is likely to find that she is herself under certain 'constraints', and that one of the 'rules of play' is that if you are obliged to teach thirty-five children in a small classroom, it is probable that there are some goals which you may not attain. The good teacher, however, does not disregard theory, but absorbs it into her practice as far as her convictions and the conditions of her teaching allow. If this were not so, there would be no ripple of forward movement in pedagogical techniques.[3]

## NOTES

1 J. D. Williams, *Teaching Technique in Primary Maths* (National Foundation for Educational Research in England and Wales, 1971), p. 69. A clear brief statement of Dienes's general approach is given in this most useful little book, pp. 65–70.

2 Zoltan P. Dienes, *The Six Stages in the Process of Learning Mathematics* (Paris: OCDL, 1970). Published in English translation (trans. P. L. Seaborne), NFER, 1973. All the quotations concerning the six stages are taken from this booklet, pp. 6–9. The reader will recognise that the later stages are applicable in the infant school only in a very simplified form. Nevertheless, this does not destroy their inherent validity, *in principle*, in regard to the mathematical learning of quite young children.

3 For further useful information on Dienes's arguments, materials, etc., see K. Lovell, *The Growth of Basic Mathematical and Scientific Concepts in Children* (London: ULP, third impression, 1971), indexed references. For a variety of ideas on mathematical games (from which the teacher will wish to be selective, for not all are practicable in classroom conditions) see: Zoltan P. Dienes, *The Six Stages in the Process of Learning Mathematics* (Paris: OCDL, 1970), published in English translation (trans. P. L. Seaborne), NFER, 1973; Michael Holt and Zoltan Dienes, *Let's Play Maths* (Penguin, 1973); Zoltan P. Dienes and E. W. Golding, *Learning Logic, Logical Games* (Harlow, Essex: ESA, fourth impression, 1970).

*Chapter 4*

# Jean Piaget

Some of the work of Piaget is more familiar, even to the inexperienced teacher, than is that of Dienes. Despite this, however, Piaget has done so much over such a long period of time that any attempt to summarise briefly his research and his writings would be superficial and possibly misleading. A detailed study of his work, however, is quite outside the compass of a book of this kind, and the reader will appreciate that further reference to books of a more specialised nature is essential if justice is to be done to Piagetian theory. No more will therefore be attempted than to touch upon some of the issues which have a direct bearing on teaching techniques, and which give some indication of the contribution that this outstanding man has made to the theory illuminating children's learning. In addition, frequent reference will be made in later chapters to Piaget's findings, because it is to these that we still look so often for guidance on the teaching of mathematics to younger children.

The revolution in teaching methods brought about by the absorption of Piagetian theory into classroom practice is already well advanced. This is probably due to its child-centred emphasis and to the fact that Piaget's teaching is, essentially, that of a developmental psychologist. That is to say, his extensive studies of children's learning make an unmistakable connection between the child's development and his capacity to undertake, and profit by, certain forms of learning. This has sometimes led, in the past, to somewhat extreme views about 'readiness' to learn, and it has quite often been implied that until a child reaches a certain stage of readiness, for example to read or to form particular mathematical concepts, it is pointless and probably damaging to try to teach him to do so. As a result, teachers have found themselves waiting for the magic stage of 'readiness' to arrive before attempting to introduce children to the learning for which this readiness has been thought to be a prerequisite.

In consequence of this view, children were (and sometimes still are) left to engage in pre-reading or pre-mathematical activity while readiness was awaited, thus unnecessarily hindering their progress. Unfortunately, this has too often been done in the name of Piaget and his 'stages'. The criticism of this view is that to extract only one part of a theoretical argument and ignore the rest, or to isolate theoretical findings in education from the ways in which intelligent teaching methods may make use of them, is to devalue the theory. And the theory in regard to the Piagetian stages has been devalued. The existence of stages is not really denied. But the devaluation arises from the assumption – which has been very widespread – that nothing can be done by the teacher to help the child towards readiness for the next stage; or, to use Lovell's accurately expressive phrase, to *precipitate learning*.

The point in regard to readiness is forcefully stated by Jerome Bruner:

'We begin with the hypothesis that any subject can be taught effectively in some intellectually honest form to any child at any stage of development. . . . The task of teaching a subject at any particular age is one of *representing the structure of that subject in terms of the child's way of viewing things*. . . . Thus instruction . . . even at the elementary level, need not follow slavishly the natural course of cognitive development in the child. It can also *lead* intellectual development by providing challenging but usable opportunities for the child to forge ahead in his development.'[1]

Let us then, in this context, look briefly at part of the Piagetian developmental view, remembering it is central to Piaget's thesis that thought arises from action, and that this process begins long before the child ever comes to school. This is an aspect of developmental psychology of which the teacher of the youngest children in school must be aware. The very young child at home explores and plays and performs actions with various objects. To begin with he does not know what the result will be when he performs these actions; then, as he grows experientially, intellectually, physically, and so on, he reaches the point when he can *anticipate* what will happen when he takes a certain course of action. For example, a baby is playing on the floor with some toys, one of which is a ball. By chance he pushes the ball, and it rolls. He did not know it was going to roll when he pushed it. This could be because of lack of experience with a ball and its behaviour; because he was not, before, sufficiently developed physically to be in a position to push; because he was timid, and emotionally and socially not yet confident enough to have a go; because, in fact, he was still at that stage of development when pushing a ball was not for him.

The time comes, however, when he can, and does, push a ball – and it rolls. Given the opportunity, he will before long reach the stage when he knows that *if* he pushes the ball it *will* roll. This indicates the development of thought. Lovell clearly explains the view that:

'. . . all *thought* is dependent upon actions. By *thought* we mean a *connected flow of ideas directed towards some definite end or purpose* . . . [the child] can represent to himself the results of his own actions before they occur. This is the beginning of true thought, since actions have became "internalised".'[2]

Eventually, the child is able not only to *anticipate* the result of his actions, but also to reverse the process and think himself back to the point at which he began. At first, this is at a very simple level. 'The ball rolled because I pushed it.' And it can be pushed back to him if somebody on the receiving end will oblige. At a very much later stage, however, an understanding of the principle of reversibility enables the child to recognise that even if he changes the arrangement or the appearance or the shape of some material with which he is dealing, he does not alter its quantitative property. For example, if he partitions a set of ten red and yellow beads into two sub-sets of red ones and yellow ones, he does not alter the quantitative property of the original set. There are still ten of them in all, and the original set of beads can be restored without any quantitative change having taken place. Similarly, if he rolls a long thin strip of plasticine into a small ball, he has not reduced the amount of plasticine even though it now looks less. The process can be reversed, and the plasticine returned to its original shape, again without any quantitative change having taken place. When the child reaches this level of understanding, he has grasped the principles of 'reversibility' and 'conservation', principles which play a most important part in signifying the stage of his mathematical development.

'The fundamental skill that underlies all mathematical and logical (internally consistent) thinking is the capacity for "reversibility"; i.e. the permanent possibility of returning in thought to one's starting point.'[3]

Piaget holds that this capacity originates very early in life, and its development and that of the child's thought processes are profoundly influenced by the way in which he is able to adapt to his physical environment. As he acquires strategies which enable him to come to terms with, and even exercise some degree of control over, his environment, he constructs a learning system. Piaget calls these strategies *schemata* or *schemes*, and the quality of the child's schemata will influence the quality of his learning system. The quality of the schemata themselves may be affected by genetic factors, and certainly experience and environment will play a most important part.

These general principles have very specific application to teaching and to the formation of mathematical concepts. The child may be provided with a variety of concrete materials, and by internalising the results of his actions with these materials he engages in thought processes that enable him to abstract the materials' mathematical characteristics. But the provision of the materials themselves (i.e. the environment, or part of it) is not enough. The quality of the strategies that the child uses to adapt to the environment can undoubtedly be influenced by the way in which the teacher helps him with his strategies. This, surely, is what teaching is about, and it relates directly to much of the content of Dienes's argument about experience, abstraction, representation, language and, finally, proof.[4]

It is against this background that we should now return to a reminder of Piaget's stages of development. There are, broadly, three:[5]

1. *The Sensory-Motor Phase* (The first 1½–2 years of life)
This is the stage during which the child's rudimentary schemata develop to the point when he recognises objects and expects them to stay there even when he is not looking; knows familiar people and realises that if they are out of the room they have not ceased to exist; and establishes certain connections between cause and effect: for example, he shakes his rattle and it makes a noise. By the end of this stage these early schemata are well developed and quite extensive, although the thinking behind them is still very vague and uncertain.

2. *The Concrete-Operational Phase* (1½–2 years to 11 years)
(a) *Intuitive Thought* (1½–2 years to 4–5 years). The child's world is rapidly expanding, largely as a result of imaginative play, experience, language and interaction with people and things. Many of his actions are internalised, and thought is much more conscious and directed. But everything is concerned with what is happening at the moment, and 'his thinking cannot move away from present situations without losing itself'.[6]

(b) *Pre-operational Thought* (4–5 years to about 7 years). This is the time when children are establishing firm basic concepts and are developing the judgement to use them operationally. They are moving towards an understanding of *conservation,* though we must remember that children appear to grasp conservation earlier in some areas of mathematics than in others. By the time they are about 7, however, they may recognise that certain elements of distance, length, number, mass, etc. remain constant or invariable irrespective of the manner of their arrangement and often of their appearance.

(c) *Concrete-Operational Thought* (roughly 7 to 11 years). Although the child still needs concrete materials and experience, the basic mathematical concepts are secure and they provide a firm background against which more complex mathematical operations may be undertaken.

3. *The Formal-Operational Phase* (from about 11 years)
During this time the child reaches the stage when he can move away

from concrete experiences and undertake mathematical operations in the abstract.

It would seem that, by definition, we are likely to be mainly concerned in the infant school with the stage of pre-operational thought (and, in fact, a bit beyond). However, we should beware of adhering too rigidly to chronological age labelling. Piaget's ages are intended to give only a very general indication of his stages and, if anything, teachers may find that children tend to reach these stages at an earlier age than Piaget suggests. It is more helpful to think in terms of *levels of development* and, leaving chronological age aside, to recognise that the existence of developmental stages along broadly Piagetian lines is still widely accepted. What is no longer accepted (and one wonders whether, given that the child has good teaching, Piaget would defend so narrow an interpretation) is that these stages are dependent upon development alone and that progress from one to another is not susceptible to being speeded up or slowed down in accordance with the quality of teaching given to the child. The 'readiness' argument now takes account not only of the developmental stage, but of the professional skill of the teacher as well. This however, may be less a denial of Piaget's view than an extension, or intelligent pedagogical interpretation, of it. Indeed, it provides a good example of the way in which sound theory should lead ultimately to better practice.[7]

Some experimental work undertaken by Bryant and his colleagues casts a certain doubt on at least one of Piaget's conclusions. Again, this is not to discredit all Piaget's work, but to recognise its use as a stimulus for further research; and if this points to different conclusions, the cause of children's learning is well served. It is precisely in this way that progress is made in extending our knowledge about learning processes.

Piaget, working with Inhelder, concluded that when presented with the information that A is greater than B and B is greater than C, a child is not able to infer that A is therefore greater than C until he passes the stage of pre-operational thought at about the age of 7.[8] In other words, given the information about A, B and C,

'. . . the child is unable to coordinate the first two, separate items of information in order to reach the correct inferential conclusion about A and C. If true, this claim has important educational implications, for a child who cannot combine this information must also be unable to understand the most elementary principles of measurement.'[9]

However, the results of Bryant's investigations with children between the ages of four and seven indicate that it is not the lack of the *capacity* to make such an inference (due to his being at too early a stage of development) that prevents the child from being able to do so. It is because his *memory* lets him down. By the time he comes to deal with A and C, he has forgotten the earlier information about AB and BC. Bryant found that if the child were really taught the earlier items of information in such a way that he could retain them, he was capable of drawing the correct inferential conclusion 'extremely effectively, but not perfectly'.

'The experiments demonstrate that 4 year old children can make transitive inferences about quantity, provided that they can remember the items of information which they are asked to combine. . . . They can combine separate quantity judgments very well and they can do so at a far younger age than has generally been assumed. This is a conclusion which has practical as well as theoretical importance.'[10]

The significance of Bryant's argument is that memory seems to play a much more important part than has been supposed in the child's capacity to grasp mathematical principles; and that the retention of information is itself responsive to teaching. This, again, does not necessarily deny the existence of stages of a Piagetian character; but it provides further evidence for the view that many other factors besides maturation contribute to the child's progress from one stage to the next.

Bryant's proposition raises some interesting questions for the practitioner. If memory can be helped by teaching, and if the ability to grasp mathematical principles is influenced (to the extent that he suggests) by memory, how far can we in fact go in carrying this hypothesis to its logical conclusion? Can we, for example, *teach* the child to grasp conservation? What emphasis are we to place on memory training? Does experience play such an important part after all? And anyway, to what extent is the proposition applicable in ordinary classroom conditions? Indeed, there are many more questions that come quickly to mind.

The first point to be made is that it is only too easy for anyone to oversimplify the implications of research findings and to draw superficial conclusions which would not stand the test of serious argument. Surely the way to regard information of this kind is as a clue, or one more piece in the jigsaw. This piece of the jigsaw suggests that we may be able to assist children towards mathematical understanding by helping them to remember certain essential items of information which they need to use when making logical inferences. For example, it is probably helpful for the child to remember that two and three are five in making the inference that five minus three therefore leaves two. But there is a critical difference between something which has been learned in the first place through experience and with understanding, and something which has been learned, without understanding, by rote. Bryant's conclusions do not change this. But they do suggest that, having understood, the child *should be helped to remember*, and this has important teaching implications. For one thing, it begins to attack sacred cows such as the disapproval of practice for purposes of consolidation; practising, for example, number bonds by doing sum cards in order to fix them once they have been understood.

It cannot be too often stated, however, that practice without understanding fixes little that is of very much use. And other things, which are less obvious, also help retention; the presentation, for example, of the same mathematical principle in a number of different ways. In a sense this is another form of practice, but it helps to teach the application of the principle as well. Therefore we return yet again to the view that the child is entitled to more than simply a rich learning environment in which the arrival of a developmental stage of understanding is hopefully awaited. Bryant's conclusions emphatically do not imply a return to rote learning, but they certainly add weight to the argument that learning may be precipitated, and teaching counts. In this sense, we believe, conservation can be 'taught'; but not as a fact, incautiously, and without the necessary foundation of experience and understanding.

## NOTES

1 Jerome S. Bruner, *Beyond the Information Given*, pp. 413–17; published in the USA in 1973 and in Britain by George Allen & Unwin, 1974. The italics in the quotation are the writer's, not Bruner's.
2 K. Lovell, *The Growth of Basic Mathematical and Scientific Concepts in Children* (London: ULP, third impression, 1971), p. 18.
3 Ibid., p. 18.
4 Confusion should not arise from the use of the word 'stage' in connection with the work of both Piaget and Dienes. In both cases, the word merely describes an order of learning, and their orders of learning are not mutually exclusive. They are different, because they are concerned with stages of different kinds. Piaget's stages relate to the child's total (including mathematical) development, and extend over a long period of time. Dienes's stages, on the other hand, are specific to the process of learning a 'formal' system of mathematics. But examination will show that neither denies the validity of the stages proposed by the other, and in fact they are, in essence, mutually supportive.
5 The terms used vary slightly, with different translators of Piaget's work.
6 Nathan Isaacs, *The Growth of Understanding in the Young Child* (London: Ward Lock Educational, 1961), p. 13.
7 References for further reading are included in the bibliography, but to fill out this chapter the reader is particularly recommended to: R. M. Beard, *An Outline of Piaget's Developmental Psychology* (London: Routledge & Kegan Paul, 1969); Nathan Isaacs, *The Growth of Understanding in the Young Child, op. cit.*; Nathan Isaacs, *New Light on Children's Ideas of Number* (London: Ward Lock Educational, 1960); K. Lovell, *The Growth of Basic Mathematical and Scientific Concepts in Children, op. cit.* (This

book provides clear analyses of a great deal of the work that has been done in connection with children's mathematical learning, and Lovell's own investigations give the teacher additional guidance which is most helpful in classroom practice. Frequent reference will be made to it in the following pages.)

8 But bear in mind the reservation concerning chronological ages expressed above.

9 P. E. Bryant and T. Trabasso, 'Transitive Inferences and Memory in Young Children', in *Nature*, vol. 232 (August 1971).

10 Ibid.

*Chapter 5*

# The Practical Implications

So far in this book some theoretical propositions have been made, which imply particular courses of action and the provision of certain resources for the teaching of mathematics in the early years of school. As an *aide-memoire* these theoretical propositions are now assembled in summary, and their overall practical implications briefly outlined. More detailed suggestions for classroom practice follow in later chapters.

*1. Different approaches may be more suitable for different children, or for teaching different aspects of mathematics.*
So variety of provision is the keynote: structural apparatus, simulated real-life situations, opportunities for the development of language, visual and auditory representation of familiar daily affairs, every sort of practical activity that will encourage mathematical understanding.

*2. Thought, which is a connected flow of ideas directed towards some definite end or purpose, arises from action.* (Lovell, Piaget)
'Doing' is therefore essential if constructive thinking is to take place; hence the continuing significance of 'I do, and I understand'.

*3. The child needs the kind of experience which will specifically help him to abstract, from the environment, generalisations that lead him to form certain concepts.* (Lovell, Dienes)
This means giving him plenty of opportunity of meeting the same mathematical situation in many different guises, in order that he may abstract a generalisation and finally conceptualise the intended mathematical principle.

For the abstraction to become fixed, however, the child must be able to represent it, visually or orally (Dienes). Recording of some kind is therefore necessary.

*4. Some early mathematical concepts, such as number value, seem to be much more difficult for many children to grasp than used to be thought.*
It follows that children must be provided with as much pre-number activity as, individually, they need.

*5. The order in which the teacher presents mathematical processes to the child is most important.*
Mathematics is a sequential subject, which is the main reason why it

cannot be left entirely to topic work or be fully integrated with other activity. It needs a carefully planned system of progression if understanding is to be secure.

One example of the importance of the order of presentation is that symbolic representation in which numbers are used should *follow* the mathematical understanding of number rather than precede it (Dienes). Again, this requirement will be met by the provision of sufficient pre-number activity.

6. *A structured environment is a critical factor in children's learning.* (Dienes and others)
One of the operative words here is 'structured'. The environment must certainly be stimulating, and must offer the child every possible opportunity for learning. But the learning cannot be left to chance. The teacher must make sure that the opportunities are used, for the purposes she intends: to take just one example for the moment, 'attribute' or 'logic' blocks will help the child to learn the principles of logic.

Carefully devised games are valuable for furthering mathematical learning.

7. *There seem to be recognisable stages through which children pass on their way to the formation of certain mathematical concepts.* (Piaget)
The stages may well exist, but there is mounting evidence in support of the view that they are not wholly developmental. Teaching can help to precipitate progress from one stage to the next (Dienes, Lovell, Bruner, Bryant and others). The 'readiness' argument has changed to take account of this.

8. *The understanding of 'reversibility' and 'conservation' are fundamental to all mathematical thinking.* (Piaget, Lovell and others)
Until these principles are grasped the child cannot be expected to perform operations with understanding (for example, the addition and subtraction of even simple numbers). Sometimes number operations are introduced to children before the mathematics involved has any significance.

9. *Memory appears to play an important part in the comprehension of certain mathematical operations.* (Bryant)
The notion that memory is important in concept formation is not new. Lovell and others have for long underlined this. But Bryant goes further and suggests that good teaching can help a child to remember what he is taught. This has important practical implications, and it provides further evidence that stages of learning are not dependent upon maturation alone.

PART TWO

# The Beginning of Numeracy

*Chapter 6*

# The Child and His Surroundings

## THE PRE-SCHOOL FOUNDATION

When the child of five or nearly five comes to school, he brings with him a fairly considerable foundation of mathematical experience and knowledge. This varies, of course, according to such factors as the nature and quality of his pre-school life, his degree of maturity, his emotional and social resources and his intellectual endowment. But although some children when they begin school are much more advanced than others, even the least advanced has encountered mathematical situations that provide the teacher with a foundation on which to build. This is a matter of some importance, because the teacher must be able to recognise the foundation for what it is. If she assumes a firmer base than in fact exists the child's understanding will be shaky and his progress impeded. On the other hand, children who really are more advanced will lose interest if they find themselves moving backwards or even standing still. It is therefore worth examining this question in some detail, despite the fact that it is an obvious truth. It is not always easy, particularly for the inexperienced teacher, to determine quickly and fairly accurately the stage of mathematical understanding that a child has reached, yet it is essential that she should do so if later problems are to be avoided. We must therefore consider pre-school mathematical experience and the ways in which it might be manifested when the child first comes to school.

For one thing, every child has met with numbers. There may be a number on his house; he goes to the shops with his mother on a number eight bus; he has one brother, they have two sweets each, his friend has a three-wheeler. Numbers, both spoken and written, have entered into his daily life. He may well be familiar with the 'counting words' to ten or more. When being taken upstairs to bed he and his mother may have recited them as they mounted each stair. The buttons on his raincoat may have been counted as they were done up. He may know by heart familiar songs like 'One, two, three-four-five, Once I caught a fish alive'. He may, nowadays, even know some of the counting words backwards, through watching television programmes about rockets taking off: 'Four, three, two, one, ZERO!'

It is apparent, then, that the child may be familiar with some of the vocabulary of number. But the question is – does this vocabulary have any quantitative meaning? For a good many children, the teacher may find that in fact it does not. If she puts a row of buttons in front of a child and asks him to count them, he may run his finger along the row, repeating the counting words as his finger moves but making no quantitative connection between each counting word and each button. By the time his finger has reached the fourth button, his counting words may have got to seven or eight. In other words, he does not realise that each counting word carries a quantitative value that is connected with the *number* of buttons before him. In the jargon of the trade, he has not grasped one-to-one correspondence, nor does he understand that two is one more than one and three is one more than two. The counting words have only a linguistic content, not a mathematical one. The child has learned them as he learns a nursery rhyme. For all the mathematical validity they have, he may as well say 'Jack and Jill went up the hill' as he points at the buttons.

Nevertheless, the fact that he can recite the counting words in the correct order because he knows them by heart is by no means without value, even though that value is not strictly mathematical. To have memorised them gives him a useful start. To support this point, the reader is invited to indulge in the following frivolity.

Imagine that you must accommodate yourself to an unfamiliar counting system. It goes like this:

| ba, | tang, | wift, | pok, | zup . . . |
|---|---|---|---|---|
| 1 | 2 | 3 | 4 | 5 |

At your sophisticated level, you have no difficulty in doing simple computation. Look at the words which represent the numbers once more, mask them, and do these sums quickly:

tang + zup =<br>
pok − wift =<br>
ba × tang =

How did you get on? Did you get stuck? And if you did, what caused you to get stuck? It was not the computational problem, because you are an intelligent adult and you mastered such simple computation many years ago. Then it must have been the unfamiliar vocabulary. So have another look at the words and the numbers, mask them again, and say quickly the number that is 'pok'. Perhaps you are now able to remember it, or at least remember the sequence of the words so that you can tap your fingers one at a time on the table and say 'ba, tang, wift, pok – pok is 4'. Wait half an hour, without thinking of the words at all or making any effort to memorise them. Then – how many is 'wift'?

You may well find that the unfamiliar words and their sequence, which you have not memorised to the point of retention, causes you difficulty quite out of proportion to the intellectual content of the mathematics. So now try a different counting system. In this system, the words are:

| lay, | kon, | tare, | ee, | oh . . . |
|---|---|---|---|---|
| 1 | 2 | 3 | 4 | 5 |

Say the words quickly and find the mnemonic which helps you to memorise them. So it is the familiar 'Lake Ontario'! Now mask the words and the numbers, tap your fingers on the table until you reach 'ee' and say which number it is. You will have no difficulty, because the word pattern is so familiar that you already know it by heart. You are well on the way to using this strange counting system competently, because you have now memorised the words.

The 'frivolity' is not, perhaps, a wholly unprofitable diversion. It shows how much it can help a child to remember the well-known nursery rhyme that says 'one, two, three, four, five . . .'. This memorised pattern of counting words will soon be very useful. Without it, the child would start even further back.

Similarly, pre-school experience helps in all sorts of other ways apart from numbers and counting. The child has been shopping with his mother – for some children an almost daily experience. He has heard the money names many times. He may himself have bought an ice-cream, and handled coins and received change. He has tried on a new coat. It is *too small.* He needs a *bigger* size. He may have been weighed, or have seen his baby sister weighed. He has been measured, and heard his height compared with that of his elder brother. 'Mark is taller.' He has met with time-names: 'It's six o'clock, time for bed'; or the time name of a favourite television programme.

It would be tedious to go on listing all the linguistic/mathematical experience that many (though not all) children have had before they come to school; though it may be worth adding one additional note. Not only have they acquired this quite impressive number of mathematical words, but in some cases they also appear to apply them correctly in unrelated situations. 'I'm nearly five.' 'My sweets were five pence.' 'My auntie lives at number five.'

One of the things which so bedevils mathematics teaching at the early infant stage is that this fairly sophisticated vocabulary can so easily disguise a level of understanding that is very unsophisticated indeed. And even much later than this children, especially if they are pretty able, can be extraordinarily adept at picking up clues and using them to carry out mathematical exercises without understanding the mathematics of what they are doing. For example, children may multiply from a number chart and get the right answer every time, because they have memorised the clue. They can 'multiply' with little or no understanding of the mathematics of multiplication. When, later on, they have to apply the *principle* of multiplication in a different situation, the clue is not there and the fog comes down.

The purpose of underlining the mathematical validity, or lack of it, of a child's pre-school experience is not to write it off as being of no consequence, because it is quite apparent that it is. But the teacher must be able to recognise the limitations of this experience and set about helping the child to conceptualise what he has learned in genuine mathematical terms. If she is misled by the competence of the vocabulary, and it is not validated with the right kind of activity, then a vital step is missed and subsequent learning is built on sand. It is therefore essential that we should be aware of what is needed for many children before the conceptualisation of this early learning can take place.

## THE ENVIRONMENT

We will begin with the environment or, more exactly, that of the classroom, for this is where the teacher exercises the greatest influence. For the youngest children, newly or not long in school, the first essential is that the environment should be one in which they feel comfortable and secure. This means, among other things, that they must know what is expected of them. There should be no confusion either in the general atmosphere or in the detailed organisation, which could add to the bewilderment that many small children experience when the more intimate world of home is exchanged for the larger world of school. Everything possible must be done to help them to feel at ease, as soon as they can, with a way of life which to some children is very daunting.

Nothing and nobody in his school environment is as important to the child as his teacher. She is the central figure of his life for many hours of the day, and his relationship with her has more effect on his capacity to learn than any other single factor in this new situation. If the relationship is a happy one, he will try to learn because he knows she wants him to; but if it is strained, the problems for both child and teacher are greatly magnified.

Every teacher knows that it is only too easy to paint an idyllic picture about this, and to imply that it simply needs a proper awareness of what is needed for her to create an atmosphere in which everybody loves everybody else and learning gets off to a splendid start. But of course it is not quite like that. There are almost certain to be some tough little nuts who seem determined from the first to turn the engagement into a battle; some who evidently concentrate on 'developing as individuals', almost as though they were born to identify themselves with this fine piece of modern jargon; others whose misery is either openly tearful or stoically contained, and who are waiting only for the moment when this unknown adult can be thankfully exchanged once more for mum; and those, probably the majority, who accept school with varying degrees of excitement or resignation, and who are prepared to go along with it all as long as their personal resources are not put too severely to the test.

These are the children with whom the teacher must now engage in the partnership of learning; and that is not an unrealistically grand phrase, because the most valuable learning takes place when both teacher and child are on the same side. It is the teacher who has to try to make this partnership a reality. The child is too young, too vulnerable, too inexperienced to take the initiative.

In comparison with the skill and the personality and the sheer humanity of the teacher, everything else in the environment is dwarfed. Yet, in another sense, the teacher and the environment she creates are complementary. They operate together to give meaning to the children's school lives. A good relationship cannot flourish indefinitely in an impoverished learning scene, but it is the teacher who causes the scene to have vitality and purpose.

The child's physical surroundings will make a deep impression on him. If his home circumstances are favourable, he will expect at least as much to capture his enthusiasm when he comes to school. If home has less to absorb him, then school may be even more valued because of the new interests which are awakened. In the environment provided by the teacher there are three factors which have a particular influence on the child's will to learn:

1. The nature of the experience and activity in which he can engage.
2. The material and equipment which can make his activity productive.
3. The organisational framework within which it is all arranged.

The detail of this will emerge as we proceed. We will begin by examining the general development of number, in the early days in school, and some of the ways in which the physical environment should be used for the benefit of those children who are still at what might be loosely described as a 'pre-numerate' stage. By no means all the children first starting school will be in this position. But some, possibly a good many, will; and for them much has to happen before the natural numbers,[1] which are central to mathematics, can find their way into the learning programme.

## NOTE

1 'Natural' numbers are 1, 2, 3, etc., as opposed to 'man made' numbers, like fractions.

*Chapter 7*

# The General Development of Number

Let us for a moment look ahead to the day when the child really understands that five is five, whichever way it is presented – when, in fact, he has grasped the principle of conservation. We will then retrace our steps, and examine the kind of learning that must take place before this level of mathematical understanding can be reached.

For clarity, let us call conservation Step 5.

*Step 4.* Before the child can assimilate the concept of the invariable fiveness of five, he must be able to abstract the quantitative property that is five, as distinct from all other quantitative properties; and he must be able to make a generalisation about this abstraction, which he can apply to five marbles or five boys or five of anything to which this common quantitative property applies.

To do so, he needs an understanding of number value, so that he can select five beads from a collection. He does not yet realise that they are *always* five, irrespective of size or arrangement, but he does realise that he is making a particular selection which is different from the selection he would make if he were asked for three beads.

*Step 3.* Before Step 4 can mean very much, the child must be able to recognise the numbers with which he is dealing, both spoken and written; and he must have mathematical understanding of the counting system. This means counting with one-to-one correspondence, and recognising the quantitative significance of each number being one more than its predecessor. Counting as a nursery rhyme is replaced by counting as a mathematical experience, and the child is ready to attach symbols to the counting names.

*Step 2.* Matching one bead to one counting name, as in Step 3, presupposes an understanding of the process of matching; matching a cup to a saucer, a bottle of milk to each child, a Unifix cube to each of three ducks drawn on a card. The teacher can help to introduce the association with counting by saying the counting names conversationally on appropriate occasions.

*Step 1.* A prerequisite of the process of matching is the ability to engage in the logical ordering of concrete materials; sorting them into collections which have common properties, placing them in order of size, finding the biggest and the smallest, and so on. Clearly pre-number experience encompasses a great deal of activity that has nothing directly to do with numbers at all.

We will bear in mind that these steps are not in practice so separate and distinct that no part of any one step is encountered in any form at all until that step is reached. For example, the child's pre-school foundation of number language will ensure that he uses numbers in speech before he meets them 'officially' in a mathematical context. The steps merge, without clear demarcation; but, with this reservation, we may now summarise them in order:

1. Sorting.
2. Matching.
3. Number recognition and counting with mathematical understanding.
4. Number value and the abstraction and generalisation of the quantitative property of numbers.
5. Conservation.

We noted in the first chapter that teaching in the modern sense includes the provision by the teacher of the right kind of activity and materials and experience, at the right time and in the right order, so that the child may be helped to learn what his teacher is seeking to teach him. We will now consider, in some detail, the kind of provision that she might make in order to lead children through each of the steps we have described, beginning with number.[1]

No attempt will be made to give an exhaustive list of every possible activity in which a child might engage. For one thing, this would be impractical in a comparatively short book; and in any case it would be unnecessary, even for the inexperienced teacher. We will content ourselves with examples of the kind of things children might do to help them with the *principles* of the learning involved in each step, and every teacher will add to the examples, and probably modify them, to take account of her particular interests, her teaching conditions, and the manner in which the children in her class react to the learning situations she provides. It is not a question of 'thou shalt'; but perhaps 'thou might'st', until thou findest thine own salvation![2]

STEP 1

*Sets*

One of the child's first experiences of embryonic mathematics, long before he reaches school, comes from things he plays with and which, consciously or unconsciously, he encounters in collections. It may be a collection of several Dinky toys, or a 'collection' of the two shoes he is wearing, or a collection of wooden building blocks. And a collection, after all, is simply a set, which to begin with has no numerical significance. But according to Fletcher:

'A fundamental and unifying idea in mathematics is the idea of a set. The children's first concepts of number come from their experience with sets of objects. From their activities in comparing, joining and partitioning sets they derive their understanding of operations with number.'[3]

This is not to imply that children should not have other kinds of mathematical activity as well; we have already discussed the need for them to engage in a variety of experience to assist concept formation. However, given that other things are also provided, the early introduction of children to sets is now regarded as of signal importance, for the following reasons:

1. From working with sets they begin to extract the pattern and structure which are fundamental to mathematical relationships.
2. Since the formation of certain concepts seems to be closely associated with identifying the common properties of members of a class, it is logical to help children who are still at an early stage of this process to do just that: to sort objects into classes by identifying their common properties, and to put them into sets.
3. Because children, from a very early age, have done things in their daily lives with collections or sets (though they have not isolated them as such), sets have the advantage of natural familiarity from the beginning.

Even for children who already have some mathematical understanding of the natural numbers (and some have, before they begin school), activities with sets are valuable for introducing number operations. By joining sets children meet the principle of addition, by partitioning them they encounter subtraction, and so on. (More will be said about this in Chapter 12.) Sets enable children to record the results of their investigations in a manner appropriate to the level of their skill. This may be quite sophisticated, or it may simply be the physical act of arrangement or re-arrangement at a time when written recording is still too laborious.

So children are given a range and variety of everyday things for sorting and classifying, in order that they may extract common properties in as many different ways as possible. There can hardly be a clearer statement of how to introduce children to sets than is given in the

Teacher's Resource Book for the Fletcher series to which reference has just been made; and whether or not the teacher is in a position to use the children's books in the series with her class, she should study the Resource Book for its wealth of practical ideas and useful information. The children's books, which are expendable, may be a little expensive for some schools when times are hard. However, if they can be provided, they help to give both teacher and children a line of advance in a way that is structured naturally though without unnecessary constraints.

One point, though, should be recognised if these books are used. The value of the approach is seriously diminished if the teacher does not arrange her organisation so that she can give a reasonable proportion of her attention to a small group of children who are working with them. For one thing, she will need to explain how the books should be used, especially at this beginning stage; and also the essence of the approach is that constructive discussion should take place between child and child and teacher and child. Without this, much of the value of the series is lost and it can become less a learning experience than an occupational one. So the teacher cannot just leave the table by the window to get on with their 'Fletcher maths' until playtime while she hears the next group read.

A piece of equipment that has brought positive enlightenment to schools in recent years is the collection of 'attribute' or 'logical' blocks devised by Dienes.[4] We recall his statement that if we want children to form logical concepts, we must provide them with materials that have more of the logical attributes than are likely to be found in the natural environment. It is for this purpose that the Logiblocs are expressly designed. There are forty-eight pieces, with a number of distinguishing attributes: three colours (red, blue and yellow); four shapes (circle, square, rectangle and triangle); two sizes; and two thicknesses. This variety of attributes enables the child to identify a considerable range of common properties, beginning with very simple identification and progressing through to much more complex operations. At first, the child might identify only one common property (all the blue pieces) and 'put a ring round them to show that it is a collection'.[5] From this he may progress to two common properties (all the round blue pieces). He might then separate the round blue pieces into thick ones and thin ones – and he partitions the set, by putting a stick or coloured lace between the two groups. He may go on to making two sets side by side, square pieces in one and red pieces in the other. He finds that some square pieces are red, so the two sets intersect with the square red pieces in both. The child's experience of finding common properties and classifying the pieces according to logical 'rules of play' is rapidly extending. Yet once again the activity loses so much of its value unless there is discussion about what the child is doing.

Before the teacher encourages the child to sort and classify, she will let him play freely with the pieces until they are quite familiar and he has recognised for himself some of their distinguishing characteristics. Conversation is needed here as well. 'What colour is that piece?' 'Do you know the name of that shape?' 'Tell me what you are going to do with them.' This helps the child to identify the pieces and to acquire the descriptive vocabulary which will assist him when he begins to undertake more structured operations. The teacher must use her judgement as to the time when she thinks he is ready to move from 'free play' to playing in accordance with some of the 'rules'.

When he progresses to this, leading questions and carefully timed suggestions from the teacher are vital if the child is to see his way into the operations that are possible with the Logiblocs. There is much to be gained from two children working together. An idea from one may be developed by the other, and there is more opportunity for constructive verbalisation than if learning is so highly 'individualised' that each child works in isolation. The advantages of interaction between children should never be sacrificed to the cause of providing totally individual programmes.[6]

*Other Structural Apparatus*

Apart from Logiblocs, structural material of a different kind is also needed in the classroom. Several varieties are available, each of which has its particular characteristics and advantages (and certain limitations). Among the most widely used in the early stages of learning mathematics are:

*Cuisenaire Rods.*[7] These consist of sets of rods, of one to ten units in length and of different colours, from the smallest which is white and is a centimetre cube to the orange rod which is ten centimetres long. The rods are not segmented, because they are designed primarily to help the child to discern mathematical relations (for example, a blue and a white are the same as an orange). The 'pure' Cuisenaire method specifically rejects segmentation, in order that numbers and counting should not intervene to inhibit the child's discovery of mathematical relationships. The rods are at first identified by colour alone, and only at a much later stage is it intended that number associations should be made.

*Stern Apparatus.*[8] This also consists of different coloured rods of from one to ten units in length, but they are segmented and there is some supplementary material, such as a number track, various trays and a board, which are designed to help the child to understand the mathematical structures he is assembling.

At the 'pre-number' stage, children should have the opportunity of becoming quite familiar with the rods. This would be achieved mainly through play which is largely undirected, but it can become rather aimless unless the teacher talks to the children from time to time about their 'play', and makes a few informal suggestions about what they might do. Children can be encouraged to experiment with the rods and see if they can put some of the shorter ones side by side to 'make them the same as' a long one. They can make a 'staircase' by arranging them in order, one above the other. They can set them out in order of length, and make comparisons which promote the use of language – *longest, shortest, the same as.* They might try to make a pattern in the lid of a box. They can certainly sort them into sets of colours, preparatory to returning them to their own box before putting them away.

Informal suggestions such as these give play a purpose while the children are becoming familiar with the rods. The children are also, incidentally, engaging in some useful learning while ordering, comparing, sorting and assembling shorter rods to match longer ones. In this way they are having really valuable preliminary experience against the time when they will be using structural apparatus 'to see the mathematics happening'.

*Unifix.*[9] This has been long established in our infant schools, but more recently the range of supplementary material has been greatly extended and it is now one of the most versatile types of structural apparatus on the market. It consists of separate interlocking cubes of ten different colours, a number track, and a great variety of trays and containers which help the child to *construct* numbers, and to see their relationships, in progressively more advanced steps as he goes along. At the early pre-number stage, children can familiarise themselves with the material by using the cubes as little building blocks and also by making patterns with the different colours. The cubes can be sorted by colour and put into sets, or 'rods' of different length assembled and set out in ascending or descending order. As with the Cuisenaire and Stern materials, comparisons can be made which encourage the use of language, with the addition of words like *more* and *less.* 'I'll put some more on and make my train longer.' The children can assemble the cubes as they would a block graph, to make a pictorial representation of something that does not need to be retained in more permanent form. For example, in the twenty minutes that remain before lunch time, a 'block graph' can be made of children wearing different coloured socks or tights. *More* people are found to be wearing blue ones today than any other colour, though nearly as *many* are wearing white. Only *one* child has green socks.

Though some of these activities are not exclusive to Unifix apparatus, one of its great merits is that it is flexible enough to be used in so many different ways. However, one kind of apparatus has some advantages, and another has others; and if the teacher is in a position to choose, her choice will depend largely on her personal view. However, if the school is already equipped with a particular kind, she should know how to help the children to make the best use of it.

There are many other kinds of structural apparatus on the market, but Cuisenaire, Stern and Unifix have been particularly mentioned because they are probably those which are most frequently found in infant schools. A certain amount of research has been done to try to determine whether one is inherently superior to another in promoting mathematical understanding or whether the use of structural material produces better results than 'conventional methods'. The most interesting fact which seems to emerge is that 'teaching quality is more important than method'.[10] This supports the view expressed in the report on mathematics in primary schools for the Mathematical Association, published in 1970:

'The manuals which accompany the various types of apparatus indicate how each is intended to be used, but they should be read very critically. Teachers must assess for themselves the contribution which such apparatus can make for their particular pupils, remembering that a variety of experience . . . is the most valuable basis for sound mathematical learning. . . . No single form of apparatus can be sufficient in itself to provide the necessary background of experience. . . . It is just as possible, and profitless, to teach children to perform tricks with numbers by the manipulation of rods as it was in the past to drill them in rote computation which they did not understand.'[11]

So we return to the view that variety of provision (of other materials as well as those which are 'structural'), *allied* to teaching quality, is what children really need; and in the child's early pre-number days, structural material should be used in the kind of informal ways which have been outlined, the purpose being to familiarise him with them in a constructive way so that activity that is more obviously 'mathematical' can beneficially be undertaken at a later stage.

*Sorting Boxes and Materials*[12]

For children to sort things into sets the teacher will want to provide a variety of materials. However, it can so easily become muddled, and its value as a vehicle of learning is then much reduced. It is often stored in something like shoe boxes; but the storage system needs to be well organised and readily recognisable if such young children are to use the material without confusion, and help to keep it in order.

The boxes should be fairly small – about the size of those for children's shoes – with not too much in each. If a child is confronted with an enormous pile of things to sort, he hardly knows where to begin.

So the teacher should be selective, and choose for the boxes a sufficient range of shapes, colours, sizes, textures, etc. to give the children variety without overwhelming them with sheer quantity. A selection could be made from such materials as buttons, bottle tops, counters, pegs, small toys like Dinky cars and farm animals, and other suitable objects. Texture and shape can be varied by sticking pieces of velvet, tweed, sandpaper, etc. on to cardboard squares, triangles and other shapes. Variations in colour and size will be included. About four of these boxes should be enough. They might be covered in Contact of the same colour or design in order to identify them (and, incidentally, to make them a good deal more durable). If this is not possible they should bear some other means of identification, like coloured stickers.

Apart from sorting these things into sets, with rings round them (as described earlier), some children may find it helpful to begin by just sorting them according to one attribute only, into trays with compartments. These trays are obtainable from educational suppliers and have a variety of uses. For example, they can be used later on for sorting collections of different numbers of items into the compartments, for storing coins of various denominations, and for holding other small equipment. If funds will not run to these trays, a good substitute can be made very cheaply by gluing trifle or cake cases into the lid of a box.

Some other sorting boxes, differently identified, can also be useful. For example, there may be one or two 'measuring boxes'. These would contain items of different length (and some of the same length), for ordering and comparing. Material like wool, ribbon and string become easily tangled and the teacher ends up by having to do more sorting than the children. It is better to stick to materials which have some rigidity, like dowelling rods or wooden beading, spills, plastic straws (which are a little stronger than the ordinary kind), or plastic-covered curtain springs cut into different lengths. There may be another box or two containing things of different size, again for ordering and comparing – for example, buttons, three-dimensional shapes such as cubes and blocks, animals, various sizes of plastic tops from aerosol containers, etc.

The teacher's aim with all this material is to provide enough variety, for different purposes, without introducing so large a quantity that it interferes with the activity of sorting; and with the variety, to store it clearly and train the children to help with keeping it all in order. Putting it back in the right boxes is itself a constructive sorting experience which benefits the teacher as well as the child!

### STEP 2

This step develops almost imperceptibly from Step 1. Indeed, it is so much an extension of the first step that it would not justify being presented separately if there were not certain matching and counting activities that are better delayed until the earlier learning is secure. These are the activities which are devised for the express purpose of preparing the child to count with one-to-one correspondence.

As we have seen, the counting names are already familiar and may be known to about ten by heart. But we have not yet concerned ourselves with introducing numbers as such, and during Step 2 we begin, informally but more deliberately, to do so. The following suggestions are examples of matching and number activities which would now be suitable.

1. When a child has sorted two sets of objects – say the yellow triangles in one ring and the red ones in another – he connects or 'matches' one from each set with a piece of string. If by now he is sufficiently skilled with a crayon, he then draws each set and is shown how to connect the objects with arrows.

2. He is given sets of pictures which have a natural connection, half the pictures being pasted on green card and half on white. For example, a bird would be on green card and a nest on white. Other pairs might be a postman and a letter, a bus and a bus driver, an egg and an egg cup, a car and a garage. The child sets the green cards out in one ring and the white cards in another. He finds the connection between each pair of pictures and joins them with string. If he can, he then draws one or two pairs in his book and connects them with arrows.

3. He has a tray with compartments and a sorting box, and he puts *one* of each of the objects in the box into each compartment. There may be a marble, a button, a conker, a bead and a peg. If time and his span of concentration allow, he can then arrange the contents of two of the compartments in two rings, and join the pairs with string.

4. He begins to use some of the commercially produced material which is designed to help him to associate numbers of objects with particular numerals. An example of this is one of the pieces of Unifix apparatus, which has plastic trays with the numerals stamped on them, separate cards depicting the numerals for matching, and the corresponding number of knobs on each tray to which the child attaches the cubes.

Another useful item consists of large wooden numerals with holes to accommodate the number of pegs which the numeral represents, and the child matches the pegs to the holes.

These are just two examples of the kind of material that is really valuable at this stage. There is a very great variety of such material on the market, but the teacher should select it with care. Some of it offers sound learning possibilities and some does not; one piece of apparatus is simple to use while another defeats the child with its complexities. In making her selection, the teacher should consider the following points:

1. What is the precise purpose of the material? Will it help the child

to learn what she wants him to learn *at that stage*? Does its *educational* value justify its purchase?

2. Is it straightforward enough for the mathematics not to be submerged by the complications of using it?

3. Are the separate pieces so small that the child has difficulty in handling them or in not losing them?

4. Is it attractive and durable?

Various incidental activities also help the process of matching. These are quite ordinary things, like putting a straw in each milk bottle, giving a bottle to each child, handing out a pencil to each one at the table. This kind of thing goes on as a matter of course, but the teacher should make sure that every child has a turn at performing these duties, for their mathematical as well as their organisational benefits.

Many of the learning experiences in Step 1 are, of course, continuing, except that some of them are gradually becoming a little more structured. The teacher, in other words, is being more specific in suggestions about how to use the Unifix cubes or the Cuisenaire rods. Although there is still an element of free play with these materials, there is also an element of directed play (like using the Unifix as described above).

Counting is another activity which is now more deliberately undertaken in preparation for Step 3. Many opportunities should be taken in odd moments for using the counting names to about ten and attaching them to objects. A group of children can be counted, the lights in the classroom, the number of milk bottles on a table, the red circles in a set, the rods in the 'staircase'. However, great care must be taken to see that whatever is being counted is actually touched or unmistakably pointed out, so that the linguistic statement gradually acquires mathematical meaning. Number rhymes and songs, with actions, are of value here; so are songs which are given some kind of concrete representation. For example, ten children facing the class can represent 'Ten Green Bottles'. The child at the end of the row sits down when a bottle 'accidentally falls', and the remainder are counted each time to see how many are left. The counting names are beginning to take on quantitative significance, along with each number being 'one more' or 'one less'. Step 3 is approaching, and it will not be long before the child is ready to move on to a more advanced level of mathematical activity.

### STEP 3

Experience and practice are needed to consolidate the learning that will take place in this step. The counting activities of Step 2 are continued and they gain more precision as the child's understanding develops. He uses more of the materials that help him to make quantitative associations between numerals and a given number of objects. He learns to recognise the numerals to about ten, both by ear and by sight.

Again, the teacher must be careful about the nature and purpose of materials and games she makes available. This is perhaps best illustrated by giving one or two examples of activities that are, at this stage, *not* suitable. Take the game of tunnel ball, for instance, in which children roll balls into numbered tunnels. This is extremely confusing to the child whose one-to-one correspondence is not yet very certain. One ball goes into one tunnel and the tunnel is labelled '4'. So the 4's and the 1's have no separate identity, and mathematical understanding is hindered rather than advanced. Tunnel ball is an excellent game later on, when scoring gives children practice in adding; but at the moment it is of no help, and indeed it can positively impede progress. The same applies to the game of throwing rings on to numbered hooks on a board. Skittles on the other hand is a suitable game, because the child can actually count those which his ball has knocked over.

When commercially made materials are being used the teacher should now select some in which the numerals appear. More learning takes place when three pegs are fitted into a board on which a *3* is depicted than into a plain pegboard, because the child is, simultaneously, counting and assimilating the association with the symbol. And once again Unifix apparatus has an advantage over other structural material in that the child can assemble the cubes one by one, and count *one more* each time.

Pictorial representation, in the form of block graphs, now really comes into its own. As the children and the teacher assemble them, sticking coloured paper or perhaps matchboxes into the appropriate columns, the principle of one-to-one correspondence is made very clear; and in the discussion which takes place during and after the assembly of the graph, there is a great deal of opportunity for counting and for making mathematical comparisons such as *two more* and *one less.*

Number symbols should be presented both orally and visually. There are many opportunities for children to see the symbols, but not quite so many for them to hear the number names except when they are counting. So they should sometimes be asked, for instance, to go fishing and bring five fish to the teacher, or to hold up three fingers at the teacher's request. They still, of course, have to count out the fish in order to collect five, but they are learning that spoken numbers as well as written ones have quantitative application. We are not yet setting out to teach number *value* as such, because this really comes in the next step; but in helping the child to recognise the written symbols, as well as their names represented in speech, we are helping him to begin assimilating the notion of value, conversationally and more or less incidentally.

STEP 4

Once the child can recognise the numerals and count confidently, with one-to-one correspondence, to ten or thereabouts, he needs plenty of experience with materials designed to lead him towards an understanding of the conservation of number. Although a great many children are likely to reach this stage much sooner than Piaget in his earlier writings proposes, there is no evidence to suggest that we can directly *teach* a child how to conserve in a theoretical sense. However, if teaching is taken to include the provision of the kind of experience that is likely to help him forward, then indeed there is a good deal that we can do.

The child must have plenty of experience of dealing with numbers and collections of objects, up to five or six at first and then to ten; putting cubes into numbered trays, threading given numbers of beads, sorting collections of objects into numbered compartments, drawing or tracing a certain number of pictures, and so on. However, activity of this kind should be presented to children in two stages:

1.

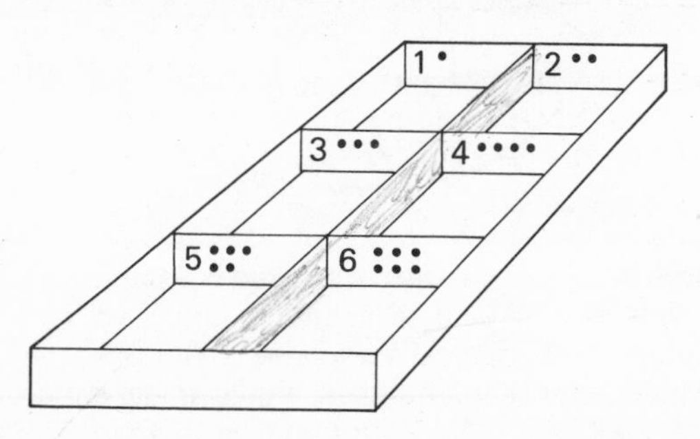


Numbered compartments for the first stage, with coloured dots that give the child a 'clue'.

2.

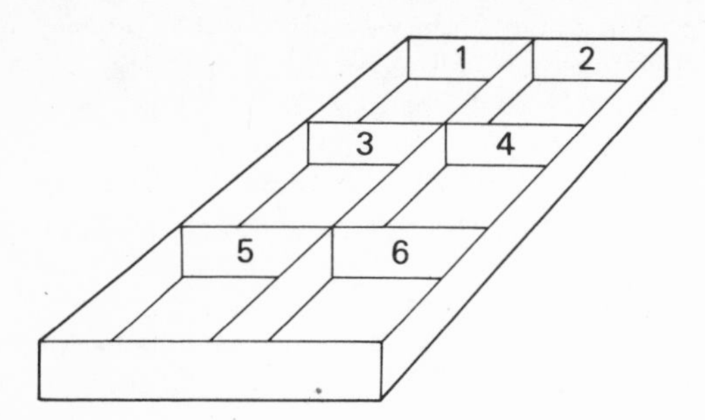


Numbered compartments for the second stage, when the child no longer needs the dots to help him.

Similarly, when threading beads the child is given some small cards showing the number of beads to be threaded, and there is a hole in the centre of the card so that it can be threaded as well.

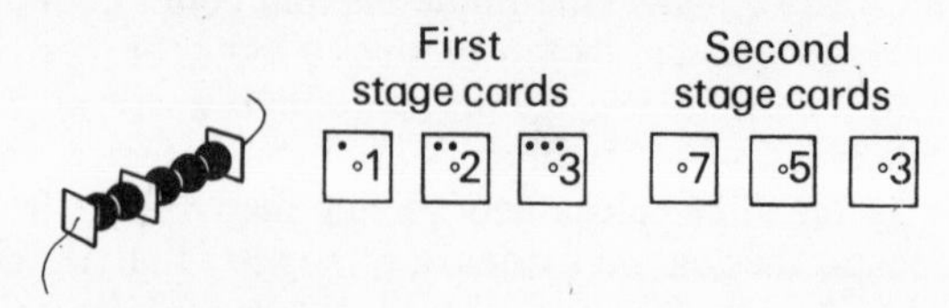


*Making sets with numbers*

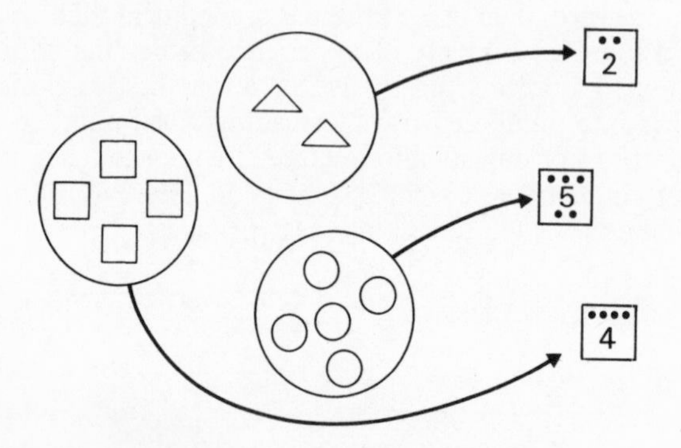


Again, the second stage cards would have no dots.

While all this is going on the children are, additionally, having a variety of other mathematical experience with shopping, measuring, etc., all of which helps them to advance their understanding of number. It is the experience of dealing with numbers in a variety of unlike situations that ultimately enables the children to *abstract* the common quantitative property of three – three pennies or three beads or the three hands that it takes to measure across the chair. The language of number emerges through discussion among the children and with the teacher and this, together with recording, helps to fix the abstraction in the child's mind. Although there is still a good deal of verbal and practical recording, children's recording on paper is increasing and this should be encouraged within the limits of their skill in controlling a crayon or a thick pencil. (More detail about recording will be given in Chapter 9.)

Children will understand number value to about ten before they can actually conserve. That is to say, they will be able to count six objects with mathematical understanding and they will recognise that six is one more than five. But they will not yet necessarily understand that six is always six, no matter how it is arranged. The teacher asks the child to count out six buttons for her, and he now does this without difficulty; but if she puts the same buttons into a pile, as he watches, and asks him how many buttons there are now, he may count them again. If he does so, he has not yet grasped the principle of conservation. He does not realise that the number of buttons remains the same even when they are arranged differently. So he needs more of the kind of experience that we have been describing before the concept of the conservation of number is firmly fixed.

In Piaget's view children pass through an intermediate stage, when they can conserve sometimes, but not always. Again, they need more number experience to establish the concept with certainty. They will conserve with the smaller numbers, perhaps up to five, before they can conserve to ten. Once conservation of number has been achieved, a whole new range of activity is open to them. The teacher should not rush this, because it is a most important step; but neither should she unnecessarily delay the advancement of the child who is ready to go on. Though some will be there, or nearly there, when they first come to school, others will still have a long way to go before there is a really strong foundation on which the next stage of mathematical understanding can rest.

## NOTES

1 There is a certain difficulty of terminology here. There is not a sharp mathematical distinction between what may properly be described as *number* and what we call *measuring* or *shopping* or any other practical mathematical activity involving numbers. However, it is important that material should be presented clearly, and we ask the reader's forbearance in accepting terminological distinctions which in practice are not so clearly separated.

2 Apart from being mentioned incidentally, teacher-made materials such as wall charts and card apparatus will be considered together in Chapter 10. It is felt that in this early stage such an arrangement will make for easier reference when looking for ideas to supplement what is available in the classroom.

3 Harold Fletcher (ed.), Teacher's Resource Book, *Mathematics for Schools*, Level I (London: Addison-Wesley Publishers Ltd, 1970), p. 2.

4 The Dienes Logiblocs are obtainable from ESA Ltd, Pinnacles, Harlow, Essex.

5 If funds will not run to hoops of the right size, reasonably durable ones can be made very cheaply from lengths of cane formed into rings and spliced tightly with string.

6 For some ideas about games that can be played with Logiblocs, see Z. P. Dienes and E. W. Golding, *Learning Logic, Logical Games* (4th impression) (Harlow, The Educational Supply Association, 1970).

7 Cuisenaire rods are obtainable from the Cuisenaire Company Ltd, 40 Silver Street, Reading, Berkshire.

8 Stern apparatus is obtainable from ESA Ltd, Pinnacles, Harlow, Essex.

9 Unifix is obtainable from Philip and Tacey Ltd, 67–9 Fulham High Street, London, SW6.

10 J. D. Williams, *Teaching Techniques in Primary Maths* (NFER, 1971), p. 29. Useful information will be found in this book about other kinds of structural apparatus, which have not been described here.

11 *Primary Mathematics; a further report* (Mathematical Association, 1970), p. 36.
12 Some of these overlap with what might be more accurately described as 'pre-measuring' activity, but for convenience sorting boxes of various kinds are included under this heading.

*Chapter 8*

# Other Mathematical Activity



The mathematical activity with which this chapter is concerned is not broken down, as it was with number, into the four separate steps. The teacher will recognise that there is a parallel pattern of progression in all areas of mathematics, so there is no need to repeat more than the outline of the pattern. Details concerning such questions as children's recording, wall charts and card apparatus are given together for all mathematical activity at this level in Chapters 9 and 10; and there is a summary chart on page 83 which may serve as a guide to general activities that are applicable in the various aspects of mathematics that we shall now consider.

## AREA AND VOLUME

The ability to conserve in area and volume does not appear to develop very early in children, though there is evidence that conservation of area precedes that of volume. It seems reasonable to assume that some children may be able to conserve area by the top half of the infant school but the conservation of volume, even in a limited sense, is not likely to be found in the infant school at all. We must therefore restrict our objectives, in this very early stage of numeracy, to no more than the introduction of the child to very general experience that is concerned with two- and three-dimensional shapes.

This general experience, however, lays useful foundations. Playing with shapes gives the child the opportunity to make patterns; to see which shapes fit together and which do not; to compare and match and order; to learn that certain shapes are better for filling some spaces than are others; to find that a particular shape can be constructed by combining some, but not all, others; and to learn the language he will need for the activities in which he will engage much later on.

At first, most of this foundation experience will be gained from undirected (though not unorganised) play with materials such as a bag of assorted building bricks: a garage has to be of a certain size to admit a toy car, though the bus needs a bigger one; bricks of one shape may be suitable for the walls, but others are better for the roof. Similar experience comes from junk modelling, from combining different shapes for the various parts of a castle or an aeroplane. Mosaic tiles can be arranged to make patterns; the squares will fit together (or tessellate), the circles will not. Wooden jigsaw puzzles with large, easily handled pieces show the kind of shape that is needed to fit into the empty space. Constructional toys encourage experimentation in assembling a variety of shapes. From activities of this kind children learn not only about the properties of shapes; they also learn a language for describing them.

The shapes with which they begin are usually squares, rectangles, triangles and circles, though words like *cylinder* and *cone* sometimes begin finding their way into the vocabulary even at this early stage. Once again, however, for the play to be constructive and for the child's vocabulary to be extended, play should not continue indefinitely to be undirected all the time. After the beginning 'free play' stage there is a place for helping the child with ideas, for making a suggestion here and there, for displaying an engine that the teacher has made from junk and talking about the shapes that she used. In practice it quite often needs a little help from the teacher to spark off an idea about a way of using the materials, and further possibilities may then be developed by the children themselves.

Opportunities should be made for encouraging language development and for enabling children to describe what they are doing and why they think this shape is better here than that one. At this stage it is entirely by verbal means that 'discoveries' can be recorded, ideas interchanged and vocabulary extended; and although it is difficult for the teacher to find the time for all the discussion she feels to be necessary she must try to make opportunities for talking with the children, brief though the conversation may sometimes be.

The full development of this kind of activity takes us into our next stage of numeracy, but it is suggested in some detail here in order to avoid unnecessary repetition. The nature of the activity appropriate to advancing levels of development is not so sharply defined that materials must at once be used in quite different ways as the children progress. It is more a matter of the *development* of the activity and of the skill with which it is undertaken than of identifiable change; and of the gradual movement away from fairly undirected play towards rather more structured situations devised for particular learning purposes.

### MEASUREMENT OF LENGTH

To understand the principle of the measurement of length, the child must ultimately be able to construct a mental image of the breakdown of the total length into smaller equal parts, which are then combined to make the measurement. To measure length also requires an understanding of the process of iteration, or of repeating the unit of measurement in order to arrive at the end result. And even before he can approach this, he must be able to count the number of units as he repeats them.

A further requirement is that the child must be able to understand that the length of a rigid object does not change because its position changes nor, if the object is removed, does the length of the space which it occupied change because the object is no longer there. Much learning and experience must take place before the child can conceptualise these principles; and, as before, we begin by building on pre-school learning, by consolidating it, and by extending it through the activities we provide.

The pre-school learning has been largely concerned with comparisons – things are longer, taller, bigger, smaller. More systematic comparison now helps to reinforce this and to take it further; hence the purpose of sorting and ordering objects of different length, and of constructing *long* trains and *tall* towers with bricks. Sometimes, however, we may overlook the fact that some of the language of length has meaning only in clearly defined circumstances. A child may be asked to 'draw a tall house'. But how tall is tall? Tall compared with what? How can the child define 'tallness' – or length or whatever – without a basis of comparison which means something to him?

A useful informal activity is to ask a group of children to imagine that they are beetles, and tell you some things which to beetles would seem tall; or that they were trees, and say which things would seem small. You have given them a recognisable basis of comparison within which they can define, intuitively, the abstract property of 'tallness'. 'Draw a worm that is longer than your hand' has mathematical meaning, whereas 'draw a long worm' has not.

Since, later on, the first units of measurement that children are likely to use are parts of their bodies, it is helpful at this stage for them to have preliminary experience of comparing these 'units'. At present, however, much of this activity can be recorded only by verbal description. A group of children can compare the length of their paces, and of the length from finger-tip to finger-tip when their arms are outstretched. They can draw round their shoes or their feet or handspans; the drawings are cut out and their length compared with each other. The teacher or helper may need to assist with the cutting because some of these children are not yet very competent at manipulating scissors. But this does not detract from the mathematical value of the exercise, which is the comparison of length and not the use of the scissors. The drawings are then ordered and assembled and from the verbal interchange that follows the differences and similarities in length are observed, extracted and discussed.

Similarly, heights can be compared, with one group of children being arranged in height order by another group. A 'heights chart' can introduce a tangible form of recording. This will not yet be used to measure the children's heights in units of any kind, but simply to compare heights. Each child in the group stands against a paper pinned to the wall, and the paper is marked and labelled with his name. Height differences are compared and it will not be long before a child will say he is 'that much' taller than Gary. Interest can be added if the teacher can draw something like the outline of a tree or a giraffe and children's heights compared with that of 'Joseph the Giraffe'.

It is very evident that much of the mathematical value of this kind of activity arises from the discussion that takes place during and after it, and not simply from the activity itself. We remember Dienes's insistence on the need for recording and the development of the necessary language as prerequisites of concept formation and the establishment of a 'formal system' of learning. We see here the application of the principle in a way that has meaning for the child at this initial stage.

## SUBSTANCE OR MASS, AND LIQUID MEASURE

Substance or mass is defined by Lovell as the 'amount' or 'quantity' of matter, or the 'amount of material'.[1] Now if we are going to compare the mass of one body with the mass of another, we cannot confine ourselves only to the size or volume of the body – that is, to the amount of air that its bulk displaces. The comparison must also take account of the density of the bodies, and the formalisation of this concept is far beyond the understanding of the child at any time in the infant school.

We should therefore define what we mean when we refer to the concept of mass or substance in relation to the learning of young children. We mean the recognition by the child that a certain quantity of a particular substance – for example, a lump of clay – *is* that same quantity, irrespective of what he does with it. We might describe this as the introduction to a much more complex concept, which the child will not meet until long after he has left the infant school. It is, however, a most essential introduction, though it stops a long way short of any kind of comparison being made between the measurement of the mass of one material and that of another.

Here again we turn largely to Piaget for help. No one before him investigated this area of children's learning quite as extensively, and no one since has substantially modified the conclusions to which his evidence has led.[2]

It was Piaget who concluded that children are usually much later than was previously thought in reaching the stage when they can really understand that a given amount of material, whether solid or liquid, remains the same despite changes in shape or arrangement, or in the case of liquids changes in the shape of the container. We are familiar with the well-known experiments in which children were presented with water in a shallow, wide glass and watched while the same water was poured into a tall, narrow glass; and because the water level was then higher, the children at an early developmental stage said it was more. The same thing happened when they were shown a lump of plasticine, which they thought became more when it was rolled into a long sausage than when it was in quite a small ball.

The importance of this is that to children who have not yet grasped the concept of the conservation of substance (in our somewhat primitive sense), the amount of material *is* bigger or more, because a different arrangement apparently makes it *look* bigger or more. Even though they may watch the same quantity being rearranged, to them the *actual amount* changes in consequence of a change in some outward aspect of its appearance; and that, as far as the child is concerned, is that!

We do not really know how much we can teach children to understand the principle of conservation in this area, and how much is simply a matter of experience and maturation. However, to the extent that experience is important we can put in the child's way helpful and constructive experience, and this surely comes within the definition of teaching. By giving children 'play' materials such as sand (wet and dry), water, plasticine and clay, by discussing with them what they are doing, by asking the kind of questions which will help them to consider, as far as their capacity allows, the effects of some of their actions with the material – with this assistance we may help them to learn.

We should probably also give them some discontinuous material for pouring from one container into another – for example, dried beans or peas. Containers of different shapes and sizes are needed, including moulds for wet sand, but anything made of glass should be avoided. Clear plastic enables the child to see the level of the material, and is safer. Funnels make pouring easier, and waterproof aprons and cuffs help to keep clothes dry. A cloth and a floor mop should be available for clearing up. Even very young children in school can be trained to help with this, and in due course they can do it themselves. But it takes them a bit of time, and the teacher must allow for this when bringing the activity to a close.

Experience of pouring that which is pourable from one container to another, of re-shaping a lump of plasticine or clay, of discussing and verbally recording the effects of their actions – all this should ultimately lead children to the point at which they can grasp the concept of the conservation of substance. This is a continuing process, and will not all happen in the infant school. And in this first stage, such understanding is still too far ahead for there to be any point in asking children to measure substances, by weighing them or by using units of liquid measurement. This must wait until experience, good teaching, and no doubt maturation as well, make it possible for the child to undertake these activities with the kind of understanding which progress in mathematics needs.

## WEIGHING

The concept of the conservation of weight is more abstract than the concept of the conservation of substance (as it applies in the early school years). At least with a piece of plasticine the actual material is there for the child to handle and see, and even then he does not for some time master the basic principle of conservation. Weight, however, cannot be directly seen or handled. It is, in fact, a very abstract measurement of gravitational pull; but even in the much less sophisticated form in which it is encountered in the infant school and well beyond, the concept is still difficult for the child to grasp. He will reach the stage when he can understand what happens with a balance when he weighs one material against another or against a standard unit; but long after this, he is likely to believe that though a sheet of paper cannot be thrown to any effect, the same sheet screwed into a ball will hit his enemy with a certain force – because it has *become heavier* by being screwed up! (Density and wind resistance have, as we know, not yet entered into his thinking.)

Piaget tells us that the concept of the conservation of weight is usually established about two years later than the concept of the conservation of substance.[3] Lovell and one of his students carried out some experiments with children of junior school age which in general bear out Piaget's findings, although they indicate that true conservation of weight seems to come even later than Piaget suggests;[4] and certainly it does not develop during the infant school years at all. However, children during this time undoubtedly grasp, fairly early, the notion of *heavy* and *light*, and many of them reach the point when they can understand the principle of *balancing* including, in time, the activity of balancing objects against standard weights. So there is no reason why they should not do this in the infant school (though, of course, at a stage beyond that with which we are at the moment primarily concerned). But the kind of experience we can give them will prepare them for the operations and comparisons which a more sophisticated understanding will make possible in the future.

For the present, however, children who are at the stage with which we are dealing in this chapter should be given experience of handling and comparing heavy and light objects, in order to strengthen the pre-school foundation which has already introduced them to this experience through physical activity. They should have the opportunity of comparing, by picking them up and 'testing' them in their hands, the weight of a pencil and a bag of shells, a feather and a stone, a shoe and a building brick. They should be asked sometimes to guess which they think is lighter, or heavier, and then to check their estimate by testing the objects in their hands. They will find, by experience, that a smaller object is sometimes heavier than a larger one. A stone may be heavier than an empty cereal packet. Discussion will establish that *bigger* does not always mean *heavier*.

This kind of physical activity, which again must almost always be verbally recorded is, for some children, a necessary prerequisite to moving on to the activity of balancing or, in a sense, *arranging* heavier and lighter situations in order to make things balance. The child now needs to play with a balance and to observe its behaviour; to see that if he puts some cotton reels in one pan, it will go down and the other will go up. He may observe that if he puts some beads in the second pan, that pan will begin to come down and the one with the cotton reels will start coming up. Sometimes, after much trial and error, he finds out for himself what he must do to balance both pans. But a casual suggestion or a well-framed question from the teacher can help this learning along. 'I wonder what would happen if you put in another cotton reel?'

However, even when the child finally understands the principle of balancing, it does not mean that he automatically makes the connection between this and the notion of heavy and light. He does not necessarily recognise that he has made the two pans balance because he has so arranged the situation that neither is *heavier* or *lighter* than the other. All he may be aware of having done is to make them *balance*.

This is where more teaching can help to precipitate more learning. The child has already established, by testing them in his hands, that the stone is heavier than the pencil. So the teacher suggests that he puts them on the balance, one in each pan, to see what happens. The child continues to do this with other pairs of objects, which he has first tested physically; and so he finally abstracts the principle that connects heavy and light with the behaviour of the balance.

The next step is for the child to balance pairs of objects which are not so certainly lighter or heavier that he can be sure of the answer from his hand testing. So it is suggested to him that he carries out a test on the balance instead. All this experience leads him to a firmer understanding of the connection between weight and the action of a balance, and he is preparing himself very thoroughly for the time when he will be weighing objects against various standard units.

A word might not be out of place about the kind of balance which may be provided in an infant class. Sometimes, if money is tight, these are homemade (which is not very difficult), or the cheapest quality balances are bought. This is not really advisable. These balances tend to be very inaccurate, even to the extent that they react differently when the same items are placed in each pan. If children are really to learn about a balance from its behaviour, it should behave reliably or it can be most confusing. A good-quality balance is well worth the money, even at the expense of doing without something else which the teacher would like to have in the classroom.

## TIME

The concept of time is one of the most abstract of all; and as far as the duration or passage of time or reading the time on a clock face are concerned, the young child has little or no understanding of what these things mean. There is sometimes a tendency to start children on reading the clock face too soon. Until they can have some reasonably coherent idea, even at a very simple level, of what it is that the clock is measuring, 'telling the time' can be little more than an exercise learned by rote. A study in 1952 (Springer) of children between four and six, cited by Lovell, showed this general sequence of development:

'1. The child learns the time of activities in his daily life. But even this goes through a sequence. To such a question as "What time do you have lunch?", he replies with a descriptive term like "Afternoon". Or he describes a sequence of events, e.g. "After lunch I have a sleep and go home". Next an unreasonable time is given; then a reasonable but somewhat inaccurate time; and finally accurate time.
2. He can tell the time by the clock first by whole hours, then by halves, then by quarters.
3. He can set the hands of a clock to whole hours, half hours, quarter hours.
4. He can explain why the clock has two hands.

Springer's groups were drawn from above average socio-economic groups; in children from poorer backgrounds the sequence may be delayed.'[5]

Lovell's own research confirms the evidence that indicates how difficult it is for the child to grasp the meaning of time.

'We cannot be sure how much we can help children to develop their concept of time, nor do we know the means most likely to help them.'[6]

If this conclusion sounds a little discouraging, we can at least take comfort from its honesty![7]

The positive aspect of the evidence is that it informs us about what we must *not* do in introducing time to children in the early stages. It clearly makes no sense to present these children with activities which require any sort of abstract understanding of the passage or duration of time, or the relationship between time and speed. Nor can reading the clock face have any meaning at least until they can understand something of what the movement of the hands indicates and, at a still more basic level, of what the numbers signify. So, on a pragmatic basis, we can consider what aspects of time can be introduced at the beginning stage, which have some connection with the children's daily lives and which may serve also as a foundation for later learning.

Children have experience, for example, of:

1. The seasons.
2. Night and day.
3. The names of the days of the week.
4. Time names associated with daily happenings, though these are linguistic expressions and as yet have no association with the *meaning* of time.
5. Date names – sometimes – which are associated with important events, such as Christmas Day or the child's birthday. At present there should be no more emphasis than this on the names of the months, partly because the words themselves are rather difficult and partly because, unlike the days of the week, they do not come round often enough to be sufficiently familiar.

The kind of activities and materials which are suitable for extending this understanding and experience are:

1. *A display or 'mini-project' of the present season.* It might be winter. So the children are encouraged to look through magazines and find some winter pictures, perhaps snow scenes, bare trees, people wearing warm clothes, etc. Some of the children may bring one or two 'winter' things from home, such as picture postcards or a ball of wool or a piece of fur. These are assembled and displayed, with conversation and discussion about what we do in winter and why. This may lead to something about feeding birds to help keep them alive, and perhaps a simple bird table can be assembled outside the classroom window and the children can put food on it and talk about the birds when they come to feed. The teacher can sometimes steer the conversation towards 'when the spring comes . . .' or 'next summer we can . . .'. This helps the children to make a reasoned connection between one season and another, and some idea of the passage of time begins to take root.

2. *A similar display and discussion about day and night.* This kind of activity not only helps children to verbalise about the experience they already have, but it also takes this experience a little further by causing them to use conscious thought processes in developing an understanding about the passage of time in this particular context.

3. *A wall chart of the days of the week.* Large, clear lettering should be used, and it helps identification if the days are written in different colours. Each day the children can be asked if they know 'the name of today', and with the teacher's help they will, before too long, do so. A bit later on they can try and think of 'the name of yesterday' and 'the name of tomorrow'. There can be discussion about not coming to school

on Saturdays and Sundays, and what children do at home instead. Again, discussion of this nature helps to introduce an element of the continuity or passage of time, an understanding of which will be very necessary when activities concerned with the measurement of time are eventually undertaken.

4. *Wall charts of time names associated with daily events.* The chart should have a picture illustrating the event, a statement giving the time name, and a clock face showing the time. It is more emphatic if the time name is written in a different coloured ink.

The purpose of the clock face is simply to begin familiarising children in an informal way with the connection between time names and clocks, not to 'tell the time' in a conventional sense. Half hours should be written as '½ past 10', not as '10.30'. The numerical symbols do not as yet have mathematical meaning for the children; but if teacher and children 'read' the wall chart together, with illustration, words, symbols and clock face being pointed out as they go along, the significance of the various forms of representation on the chart is gradually assimilated, linguistically, and reading as well as mathematical foundations are being laid. Quarter hours should not be used at this stage, as they only add an unnecessary complication. Not more than two or three charts should be displayed at one time, they should be changed fairly frequently, and they should be regularly 'read' and discussed. For teachers whose artistic talents are limited, magazines and colour supplements often have pictures which make very suitable illustrations, and some of the Ladybird[8] books have good clear pictures of children coming to school, going to bed, eating meals and engaging in other familiar daily happenings.

5. *Month names.* At this stage, these are best dealt with entirely conversationally, with the sole purpose of making the children familiar with some of the words. 'It's December now. Soon it will be Christmas Day.'

'Today is Sally's birthday. It's the second of March.' When children eventually come to deal with months, it will be much easier if the words are not new; but there is no point in attaching them yet to dates, other than incidentally. Months, after all, do not even have anything to notify the change from one to another. Days at least have nights in between. So, although reference should be made to month names, they should not yet be used in association with notions of the measurement or passage of time.

## MONEY

Since most children have seen and handled coins for some time before they come to school, have heard many of the money names, have witnessed the transaction of shopping when out with their mothers and indeed have sometimes undertaken the transaction themselves, it is particularly easy to credit them with rather more understanding of money and shopping than in fact they have. One may realise that a child has not reached the stage when money, with number names attached, has any quantitative significance; and that the comparative value of coins has as yet no meaning. But it may come as something of a surprise to find that even the transaction of changing money for goods is still, to some children, no more than a physical action learned through experience.

In other words, they do not recognise that when you go shopping there is a relationship between money and the *value* of goods. If a child goes to a shop with 5p to buy sweets, he will be able to exercise a certain amount of choice among the sweets that are available for 5p; but within that choice he ends up with what the shopkeeper says he may have. He may certainly not realise that the shopkeeper has limited his choice because of anything to do with value. He has simply learned through experience that that's what shopkeepers do, and he has to accept it. The teacher may even find that some children, when they first come to school and use the play-shop in the classroom, are unwilling to part with their coins in exchange for goods. The coins, after all, are *theirs* for the moment, and they see no reason why they should give them up just because they want something from the shop! Other children, of course, have learned to accept this much, even though they do not understand much about money itself.

The natural way to begin building on this experience is to further such ability as the children may already have to recognise different coins and to attach names to them, and to provide a simulated real-life shopping situation by having a classroom shop. Unfortunately not all classrooms have enough space for a shop, in which case the teacher must restrict herself to providing coin recognition and naming activities, and delay the shopping part until a later stage when the children can use desk and 'wall' shops.[9] However, if space can be found for even the simplest kind of shop, it is worth providing for the experience that it affords.

*Coin Recognition*

Except for those children who are already at a more advanced level, this at first has nothing to do with the value of coins. It is simply a matter of securing in the child's mind the fact that there are different coins, which look different and have different names. The first activity is to let them have small sorting boxes of the kind described on page 56, which contain a few 1p, 2p and 5p coins. These may be sorted into sets, or into trays with compartments. The association of the names with the coins is entirely dependent upon conversation with the teacher, who will realise that this is still a naming activity in which number names are used without their having quantitative connections. It is not particularly helpful to apply vocabulary 'rules' to the naming of the coins. 'A penny' is fairly obvious, but the others may be 'a twopenny' (tuppenny) or 'a twopenny piece', or even 'a two p'. There seems little point in being too much of a purist over this. We accept that the use of more than one name may cause some confusion at first, but language is used in this way in normal daily life and children hear coins being given different names. This is the kind of situation in which the teacher's common sense is of more value than pedantic argument.

When the child is fairly accomplished at sorting, he can be given a card on which some gummed paper coins have been stuck, of 1p, 2p and 5p denominations. He then matches the coins he has sorted to their counterparts on the cards. At first he is likely to do this according to the variations in size and colour. But the gummed paper coins on the card will sometimes show the 'heads' and sometimes the 'tails'. The child may discover the difference for himself, or he may need a 'leading question' from the teacher to encourage him to do so. This provides the opportunity for further discussion about the coins and their different designs, which helps the child to memorise them and ultimately to distinguish them without hesitation. When the teacher judges the child to be ready for it, she can add the 10p to the coin recognition and naming activity.

In some schools there is a strong preference for children always to use real coins rather than plastic or cardboard facsimiles. The argument here is that if we are teaching children about money we should not make them use something which is not money at all; and in any case artificial coins are so unlike the real thing that it can confuse children and cause them to have to 'unlearn' when dealing with proper money out of school.

In other schools it is argued that the use of real money creates more problems than it solves. It cannot be allowed to disappear, and the teacher has to check it frequently and may be involved in the unhappy consequences of discovering that some of it has found its way into Willie's pocket. It is felt, too, that artificial coins are recognised by children as being artificial and that they do not become confused, because 'playing' with cardboard coins is a 'pretend' occupation and comes into the same category as playing with a toy car.

If the teacher has the choice, she must make up her own mind about the pros and cons of these arguments; but it is more likely that her decision will be governed by the policy of the school. However, if artificial coins are used, it really is essential for the children to be able to compare them with the genuine ones, which they can handle and talk about and finally name and distinguish, just as they do the cardboard counterparts. This is best done as a small group activity with the teacher, when both real and artificial coins can be handed round, compared and discussed and collected at the end without any fuss.

### *The Play-shop*

Assuming that there is enough space to have a shop, it will not afford full value as a learning experience unless the shopping activity is informally but fairly carefully organised. It is not enough simply to set up a shop and then let the children loose to play with it. This can soon become very purposeless, and interest in the shop is lost.

The teacher should begin by engineering a discussion situation, which arouses the children's interest in having a shop and helps them to feel that setting it up is a co-operative venture in which they can play a part. The decision then has to be made about what kind of shop it is to be, and here the teacher must be conscious of her objectives. This first shop will be simply a play-shop. Its purpose is to give the children experience of undertaking shopping transactions, of exchanging money for goods, of learning how to look after it and of becoming interested in shopping as an activity. For many of them buying and selling in the play-shop will be anything but a precise monetary transaction. Moreover, as time goes on the shop will have to be changed. No shop will withstand indefinitely the fortunes of time and the depredations of many hands on the goods that are for sale.

For these reasons, there is much to be said for stocking this first shop with the simplest and most expendable merchandise, reserving the more sophisticated varieties for later stages when the objectives of shopping have changed. We will therefore suggest a particular pattern, in order to avoid the tedium of listing endless possible alternatives. This is not, however, to imply that the teacher should not use any alternative of her preference.

One point should be made before entering upon the detail of the shop and the way in which it might be used. This concerns the pricing of the articles for sale. There are two schools of thought on this:

1. All prices in classroom shops should be the same as those in the real world of shopping outside. Otherwise we are deliberately misleading children; and we do not give them a proper understanding of money value, because when they go shopping with their mothers they find that a tin of baked beans costs something quite different from its price in the classroom shop and they must then unlearn what we have taught them.
2. If we stick rigidly to realistic pricing in the classroom, we impose such a severe limitation on the goods we may sell that it becomes rather dull and lacking in variety. This particularly applies in the early stages, when the sums of money involved have to be very small in order to keep them within the amounts that children can understand.

Here again the teacher must make up her own mind about the policy she pursues, because it cannot really be argued that there is a wholly 'right' or 'wrong' way. However, while admitting that it is a personal view, the writer has a strong preference for the second alternative in the early stage of shopping activity. The disadvantages of realistic pricing have seemed in practice to outweigh its advantages, and there is a parallel with the arguments about real and artificial money. For one thing, money value is still so remote from the understanding of these young children that they are hardly in a position to make comparisons between prices in the classroom and those in the shops outside; and in any case they are aware from the beginning that the play-shop is not a real one. Later on prices can indeed be more realistic, but for the present priority is given to variety and practicality.

Having declared this interest, we will now consider shopping activities in terms of the stated preference, and the teacher who inclines to the first alternative will adapt her arrangements accordingly. The suggestion therefore is that this first shop should be stocked with empty food packets brought from home, and the children will be invited to bring them along so that the shop may be set up.

*Assembling the Shop.* Within the limitations imposed by space, the shop should be as realistic and as attractive as possible. Occasionally the teacher may have in her classroom a commercially made shop, complete with counter and a little canopy overhead. These, however, are expensive, and an effective alternative is to use an extra table, preferably placed across a corner so that there is access from behind for 'shopkeepers'. Sugar paper can be pinned on the front and sides, perhaps chalked with a design of bricks and with a small hook on which to hang an 'open' and 'closed' notice.

As the goods arrive (and, incidentally, the teacher will have to select them most tactfully because she will be submerged with packets extorted from mothers) there should be general discussion about how the articles should be prepared and priced. Empty packets soon become battered and squashed if they are used without being strengthened. It is much better if the children, with the teacher's help, fill them with a little screwed up newspaper and seal them with sellotape. There should then be discussion as to the 'money names' that should be given to each packet. A very limited number of money names should be used at first, probably just 1p, 2p and 3p. Bearing in mind that these are only money *names*, and have nothing yet to do with value, the children should be allowed to call this one *3p* and that one *1p*, just as they wish. The teacher then, with the children watching, writes the name – i.e. the symbol 3p[10] – clearly on each packet, with a crayon or felt-tipped pen, saying the name as she writes it. The purpose of this is to help the children to begin making the association between the money name and its application to the article when they go shopping. Finally the children, in ones and twos, arrange the merchandise on the 'counter'.

This kind of introduction to the shopping activity helps to make the shop something more than just another piece of classroom furniture. Without such an introduction, it can so easily be identified in the children's minds as little more than a collection of junk that they play around with when they have nothing better to do. The 'playing' part is justifiable; but the 'nothing better to do' part can be too aimless to have any educational justification.

*Using the Shop.* The shop is now there, and the children have participated in establishing it. There is therefore more motivation in the shopping activity than there would have been had the goods just been put there as the children brought them, or if the teacher had assembled them all after school and announced the next morning that we now have a classroom shop. But, we repeat – and it is an important 'but' – it will not be of much educational value if it is just used as a spare time plaything. Certainly it is, at this stage, a *play*-shop; but not a shop for any old kind of play, which has no identifiable purpose.

Therefore we return to what its purpose really is; and, as we said, its purpose is 'to give the children experience of undertaking shopping transactions, of exchanging money for goods, of learning how to look after it and of becoming interested in shopping as an activity'. Certain organisational arrangements will help to fulfil this purpose.

1. It is an advantage to have 'shopkeepers'. This makes it more like a real shop. The shopkeepers feel very responsible and have the opportunity to be legally authoritarian (and don't we all like this from time to time); and it helps to prevent the activity from degenerating into a free-for-all. The shopkeepers can exercise the supreme responsibility of changing the notice from 'closed' to 'open'. (Since they cannot read,

it is helpful to stick a star or a picture on the 'open' side.) The shopkeepers can also, in true British tradition, insist upon an orderly queue; and apart from providing invaluable training in British customs, this helps the teacher no end in reducing pushing and shoving to a minimum. It is, in addition, excellent for promoting early training in looking after the classroom. It is the shopkeepers' job to see that all is shipshape before the shop is opened, and to restore order when it is finally closed. Perhaps most important of all, the existence of shopkeepers draws attention to the element of a *transaction* in shopping, which is absent if self-service is carried to the length of helping yourself to an unlimited quantity of goods and departing without even handing over some of your stock of pennies in exchange.

Everybody, of course, wants to be a shopkeeper, if only for the status it bestows; so the distinction must be conferred on every child in the class. It is a good idea to have two shopkeepers on duty at a time. To institute this custom at the beginning pays dividends later on, when shopping is a more regulated occupation. It helps if one of the shopkeepers is reasonably competent. To achieve this, the teacher may draw up two lists of names, which are pinned up near the shop. As far as the children are concerned, the names are listed in a completely random manner; but the teacher in her wisdom has ensured that one list is of supervisory material, while the other has the dreamers and those who have not quite caught up with what it is all about. In this way the supervisor helps the dreamer, and the dreamer gets his turn and incidentally learns the practicalities from his partner. The establishment of this arrangement is most helpful at a later stage, when the dreamers really do need the assistance of the supervisors if the shop is to work.

As each shopkeeper has had his turn his name is 'ticked' on the list, and because every child knows that his turn will come the urgent chorus of 'please may I be a shopkeeper' is avoided.

2. Not more than four children (apart from the shopkeepers) should shop at any one time, or long queues may form when the shoppers get bored with waiting and begin scrapping. Since they cannot yet count, it is a good idea to have four cut-out pictures of children stuck on card and kept by the shop, and a child must possess himself of one of these before he may go shopping.

3. The coins should be kept in small boxes or tins near the shop. Tobacco or cigar tins are about the right size and are very suitable provided the lids do not fit too tightly for the children to open. The tins should be painted in different colours or have coloured paper or Contact stuck on the top in order that children at different stages can identify the tins they should use. Those for the first stage should contain only pennies because as the children's numerical skills develop they will reach the point at which they recognise that they must give three pennies for a 3p packet; but the added complication of assembling coins of

different denominations or of having to get change will still be beyond them. Not more than six or seven pennies should be in these tins.

At first, children at the beginning stage will hand over random numbers of coins for their purchases. They do not yet know what 'three' is, let alone be able to add the cost of several items. However, taking into account the objectives of this first shop, we do not look for accuracy of payment. After all, if the child were capable of this he would not need a play-shop of the kind we are now describing. It is one of the activities that will help him to learn, not only about coins and money value but also about number.

4. The shopping activity at this stage should be entirely one of free play in the sense that there should be no teacher direction about what the children may buy. However, the organisation of the activity certainly needs teacher direction if, as we have suggested, 'free' play is not to become 'aimless' play. In addition to the points already mentioned, the teacher must ensure that every child has experience of the shopping activity at a level commensurate with the stage he has reached. He must be trained to collect his tin of coins, his shopping bag and the picture which shows that there is room for him at the shop, to make his purchases in a reasonably orderly manner and to return them to some appointed place when he has finished. The shopkeepers need to be trained to collect all these purchases when the shop is closed and to put everything back where it belongs. If there are more advanced children in the class, they can help to sort out the coins that have been used and put them back in the appropriate tins. Otherwise the teacher, or the helper if there is one, must supervise this.

The organised shopping activity does not preclude the use of the shop entirely freely at other times, provided the teacher makes arrangements for it to be kept in order. A shop that is allowed to become an untidy mess ceases to have much value as a learning experience. The goods will in any case need to be replaced from time to time, because they are not very durable.

5. As with most activity at this stage, the value of shopping is much enhanced if the teacher – and the helper – talk informally with the children about what they are doing. No other form of recording is yet possible for these children, and it is largely through discussion that a child's shopping experience leads him towards a firm understanding of coins and their value.

Towards the end of this first stage of numeracy, when the child is counting with one-to-one correspondence, can recognise the numerals, and has some notion of the value of the smaller numbers, he will make his payments with a fair degree of accuracy. He should not yet be given a list of items to buy, or be asked to record in writing what he has bought, because this involves grouping numbers together (addition) which is of no mathematical value until he can conserve number. The shop should therefore be regarded as a play-shop until the end of this first stage, when the child will be ready to move on to shopping activity of a different, and more advanced, kind.

## NOTES

1 K. Lovell, *The Growth of Basic Mathematical and Scientific Concepts in Children* (London: ULP, third impression, 1971), p. 61.
2 There is one particular exception to this, but it concerns the child at a more advanced stage in the infant school and it will be discussed in Chapter 13.
3 But for some reservation on this view, see page 138.
4 *Op. cit.*, pp. 73–7.
5 Ibid., pp. 83–4.
6 Ibid., p. 89.
7 A pithy and straightforward comment from a colleague further illuminates the point: 'One reason for the measurement of time being perhaps the most difficult of all is that until you've had it you can't measure it and when you've had it you can't check your measurement.' (L. H. Holland.)
8 Published by Wills & Hepworth Ltd, Loughborough.
9 For these, see Chapter 13.
10 This is not at variance with Dienes's view that symbolic representation should follow mathematical understanding, not precede it. The symbol here is being used simply as a graphical representation of a linguistic label, not as the representation of a numeral in the mathematical sense.

*Summary of First Steps in Mathematical Progression*

| | *Number* | *Area and Volume* | *Length* | *Mass and Liquid Measure* | *Weighing* | *Time* | *Money* |
|---|---|---|---|---|---|---|---|
| *Step 1* | Sorting or classifying into sets, etc. | Undirected play with shapes; bricks; jigsaws; junk modelling; constructional toys | Sorting Ordering Comparing | Plasticine, clay; sand and water play; containers of different shapes and sizes | Balancing | Seasons; night and day; days of the week; 'Mini-projects'; time-names | Play-shop; coin recognition and money names—1p, 2p and 5p. Undirected but organised |
| *Step 2* | Matching, leading to incidental association of objects with numerals | As above, gradually more directed ↓ | As above, gradually more directed comparing hands, feet, etc; comparative heights chart | As above, gradually more directed ↓ | As above, gradually more directed ↓ | As above ↓ | As above |
| *Step 3* | Number recognition; counting with 1-to-1 correspondence | ↓ | ↓ | ↓ | ↓ | ↓ | As above, probably add 10p to coin recognition; payments in shop more accurate |
| *Step 4* | Number value; the abstraction and generalisation of the quantitative property of numbers | | | | | | As above |
| *Step 5* | Conservation (Chapter 12) | ← See Chapter 13 → | | | | | |

*Chapter 9*

# Discussion and Recording

Throughout these pages attention has been drawn many times to the need for discussion among children and between them and the teacher if the full learning value is to be extracted from the activities in which the children are engaging. The reasons for this are:

1. Discussion enables children to fix and extend their learning, because the verbalisation of activity helps them to recognise *why* their actions have brought about certain results and this in turn uncovers further learning possibilities. (It will be remembered that Dienes placed great emphasis on this in his fourth and fifth stages.)
2. Discussion helps language development and leads to accuracy of meaning in mathematical vocabulary.
3. It is natural for the child to want to talk. Language is for him an important resource and to make use of it helps to motivate him in his learning.
4. In the early stage it is one of the few forms of recording that he can undertake easily. This aspect of discussion decreases in relative importance as the child's skill in written recording develops, though the first three reasons retain their significance throughout his school life.

There is little difficulty in recognising *why* discussion is important, and even the most inexperienced teacher will accept that it is. However, it would be less than honest to duck the next question, which is '*when* can it all be done?' It is all very well to say that more learning takes place if the teacher talks with the children about what they are doing but, with a large class, time is the scarcest commodity of all. There is never enough of the teacher's time for all the demands that are made upon it.

So it is necessary to consider constructive ways in which the teacher may be helped to extract the maximum value from the time at her disposal. However this, as a general question, applies not only to children at an early stage of learning. It applies at all stages It would therefore be better dealt with as a whole, and it will be deferred until Chapter 14. The reader may wish to turn to that chapter next, because we will now continue with the matter of children's recording at the early stage, other than that which takes place in discussion with the teacher.

Some children when they begin school still have extreme difficulty in controlling a pencil. Even those who can do so fairly easily may still have to learn to write, and most of them cannot read either. For these reasons, any written recording of mathematics that involves words will, to begin with, be virtually impossible.

Certain other forms of 'written' recording, however, will be possible from the beginning or not long after children start school. Drawing, tracing, using gummed paper pictures and making arrow diagrams are instances of this. Some examples of the kind of recording these children might manage are given below.

1. Boxes of coloured gummed outline pictures are available from educational suppliers – pictures of boys, girls, animals, birds, etc.[1] The children can make 'a set of girls' or 'a set of rabbits' by sticking the appropriate pictures on to a large sheet of kitchen paper and, if they wish, drawing a ring round them.

2. The teacher can make 'work cards', such as:

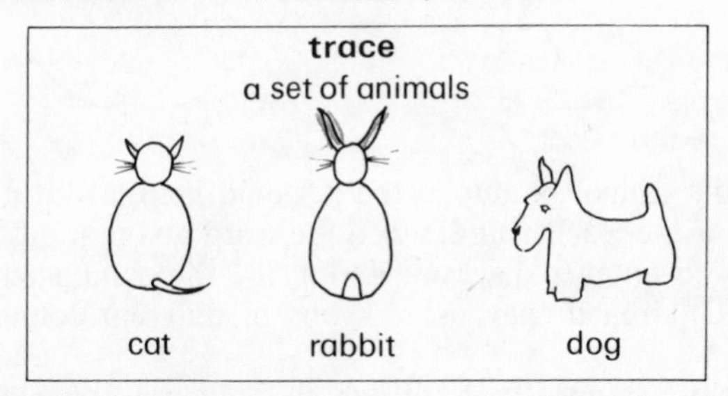


If the word 'trace' is written in red, the child learns that the card with the red word at the top is one which he traces, even though he cannot read the word. The teacher at this stage must read the other words to the child in an odd moment while he is doing the card.

3. For a change, similar cards may ask the child to draw rather than trace. The word 'draw' could be written in green.

4. *Relations* are clearly seen in arrow diagrams, though at this stage these are best done as a group activity with the teacher.

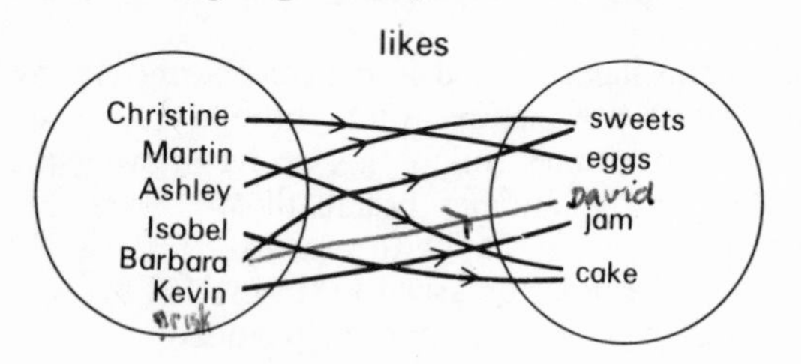

It is less confusing for the children if each arrow is drawn with a different coloured crayon. The teacher writes the words and interprets them for the children as they put in their own arrows.

Arrow diagrams are good for matching and for illustrating relations, whether they be one-to-one, many-to-one (i.e. Ashley and Barbara both like sweets), one-to-many (Isobel likes both jam and cake) or many-to-many. Naturally all these possibilities will not appear on the same diagram.

5. As children progress to counting and using numbers, they can 'draw a set of 2'.

6. If a duplicating machine is available at the school or the Teachers' Centre, the teacher can duplicate expendable sheets on which the children record early number work by means of arrow diagrams, e.g.

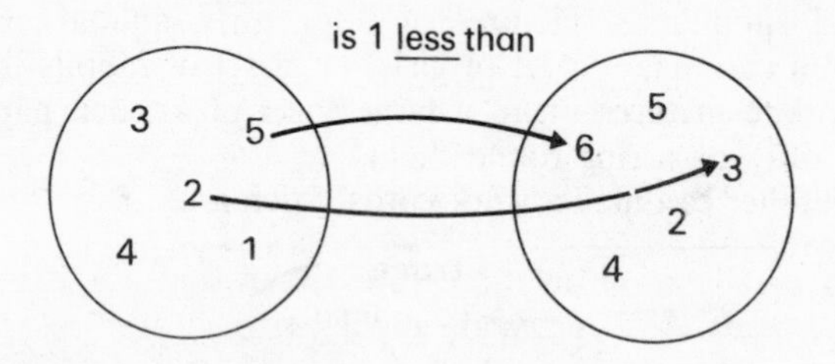


Since colours cannot be duplicated it would help, as far as reading is concerned, if the teacher underlined the word *less* in a different colour from *more* on another diagram; and again, the completed diagram is clearer to children if they use crayons of different colours for their arrows.

7. An 'old-fashioned' but still wholly valid way of giving children practice for the consolidation of number recognition and number value is for the teacher to draw different numbers of pictures in the child's book, on duplicated sheets, or on cards (see page 96, no. 11), which the child counts; and he writes the numeral beside each row of pictures. The converse activity is for the teacher to write the numerals, and the child draws the required number of pictures for each. (The teacher should not undervalue practice, which children really do need for the consolidation of learning that has come from practical experience.)

8. When a child has recorded in concrete form (e.g. by matching a '5' card to a set of five pictures, or by fitting three cubes into the '3' tray) he should sometimes follow this up by drawing the pictures or cubes and writing the number beside them. In the case of simple pictures, tracing is an alternative to drawing.

Clearly the child cannot be asked to transfer all his concrete recording to paper; but to do so sometimes helps to fix what he is learning.

*Other Useful Forms of Recording*

1. Drawing round hands and feet for comparative measurements, as described on page 67.

2. Drawing round shapes, below which the teacher writes the name of the shape and the child then copies it.

3. Tracing, and later copying the numerals. The cards from which the child traces or copies should indicate the direction in which the numeral should be formed, e.g.

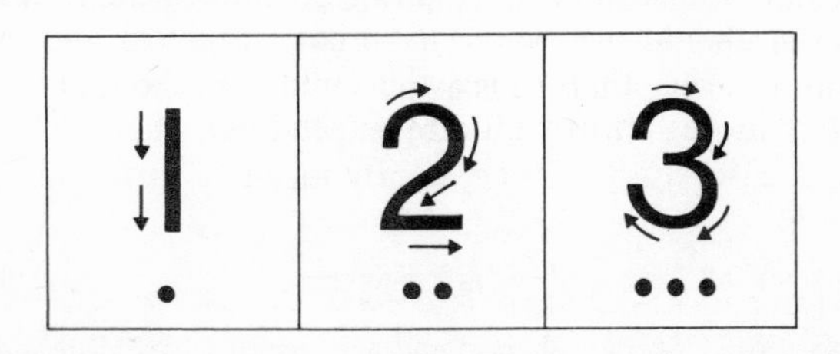

Many children will not follow the arrows at first, but with help will gradually begin to do so.

Children are often given more practice in forming letters than in forming numerals. The numerals should not be overlooked, because they are quite difficult to write and obviously it is in the child's interest to be able to write them easily.

As with letters, many children mirror the numerals. This should not be corrected in a way that makes them feel it is *wrong*. Mirror writing is so common at first, and indeed so normal. The teacher should just point out now and again that the numeral is the wrong way round, and encourage the child to do it the right way next time. Most children outgrow mirror writing with experience.

*Graphical Representation.* In addition to arrow diagrams, block graphs are widely used at this, and later stages. They really help children to 'see numbers happening', and they are invaluable for leading on to numbers and counting as well as for making use of the child's developing number sense. The teacher will remember that not all block graphs need to be assembled in permanent form on a card on the wall. Many occasions arise when 'temporary' graphs can be used to illustrate a point of discussion. Such things as bricks, square wooden beads, matchboxes, and Unifix cubes are admirable for the purpose. They may just be assembled on a table as the discussion proceeds, and dismantled afterwards.

It will be remembered, too, that a block graph is of much less value if it is not fully discussed while being assembled, and the results recorded verbally when it is complete.

T.D.



*Concrete or 'Physical' Recording.* Much recording at this early stage is of a concrete nature, and no more is necessary than to draw attention to its existence as a valid form of recording learning activities. Making sets; sorting, ordering and comparing; using the balance; sand and water play; making patterns; using rods and cubes – all these and many more activities are recorded in concrete form and often in no other form except discussion. Although the teacher will want to encourage representational recording as well, she will recognise the children's limitations in writing (and sometimes in drawing and tracing), and she should not worry if verbal, graphical and concrete recording far exceeds that which the child puts on paper until such time as he is more skilful with crayons and pencils.

## NOTE

1 These are called 'Shirley Shapes' and are supplied by Philip and Tacey Ltd, North Way, Andover, Hants.

*Chapter 10*

# Wall Charts and Card Apparatus

## WALL CHARTS

Several wall charts have been mentioned in the text so far, and for convenience these will now be listed together with others which are useful for children at this level.

*1. The days of the week.* Identification is easier if each day is written in a different coloured ink.

*2. Weather chart* – to encourage the use of time words, like *yesterday, today, this morning.*

*3. Charts of time names* (for detail see page 74); also charts illustrating the seasons, and day and night. Not more than two or three of these should be displayed at any one time.

*4. Charts of comparative sizes,* for the reinforcement of vocabulary; *big/little, tall/short,* etc; again only a limited number should be displayed at the same time.

*5. Shapes*: squares, rectangles, triangles and circles. Several shapes of different size and colour are better than a single shape to illustrate each name.

*6. A colour chart,* limited at first to the most commonly used colours. Motifs of different shades of the same colour are attractive and reflect our ordinary terminology.

*7. A heights chart.* (See page 67.)

*8. A money chart,* depicting the common coins and their names. This would not be used very much before the end of the first stage, so if wall space is at a premium it could be reserved until later.

*9. Number charts* from one to ten. These would not be used by children in the early part of the first stage, but would be needed as counting (mathematically) begins.

Number charts may have more impact if objects instead of pictures are stuck on the cards beside the numerals. The objects (or pictures) should be arranged in random patterns, or some children may believe that four is four only when it is shown in a standard way.

All ten of the number charts should not be put up at the same time. Pin them up in ones or twos, with the children participating, discussing the numerals and the numbers they represent.

Wall charts should not be allowed to recede into the classroom scenery, which nobody really notices any more. In odd moments one or two charts should be 'read', so that the children become familiar with their purpose and the words and information on them. If a chart ceases to be useful, it should be replaced by another. There will in any case be only a limited amount of wall space for charts, so the teacher should select those which she thinks would be particularly helpful at the time.

## CARD APPARATUS

Nowadays it is not universally agreed that this should be used systematically in infant schools, particularly with very young children. Certainly card apparatus needs to be given careful thought, at all stages of learning, or it can too easily become a means of occupying the child rather than of helping him to learn (though one has a great deal of sympathy with the harassed teacher of a large class who is sometimes driven to the expedient of 'occupying' children; if we are honest, most of us would admit to having done this at some time or other!).

If card apparatus is to be a genuine educational aid, it helps if attention is given to certain points of detail, some of which have special application to its use with children at this early stage of learning.

### *1. Limitation of Quantity*

Most of the mathematical learning of these children arises, essentially, from active involvement in practical situations and, as we have seen, in the recording of this learning by mainly verbal and concrete means. Much of this activity is in the nature of 'free' or unobtrusively directed play. Moreover, card apparatus is new to these children and its very unfamiliarity can sometimes present problems. Finally, they cannot sit still, working at their tables, for very long at a time, and many of them are inexperienced at doing things on their own.

It is not impossible for card apparatus to fit into this pattern, but it must be very carefully devised and there is a close restriction on the amount that is of real value. Such apparatus as is provided must be designed with the following points in mind:

(i) It should give the child a practical rather than an abstract task; that is to say, it should ask him to *do* something, rather than work something out. Towards the end of this stage he can move on to apparatus which requires him to find, for example, the number that is 'one more than four', but even then he must have counters or other concrete material with which he can undertake the task.

(ii) It should, when possible, be presented as a game or at least as a small group activity with three or four children working together. These games are not competitive, and because the children are still at a highly individualistic stage each one is 'doing his own thing'. But he is doing it with others, as he does in play, rather than by himself in isolation with little opportunity of talking about what he is doing.

(iii) It should be clear, colourful and attractively presented. Any writing should be in large letters, with words illustrated wherever possible. It should be straightforward to operate, easy to look after and accessibly stored.

(iv) It should, at least at first, be closely related to the practical activities in which the children are engaging. This helps to bridge the gap between the familiar and the unfamiliar.

(v) It should be carefully explained, so that the children are quite clear about what they are expected to do with it.

### 2. *Limitation of Its Purpose*

Card apparatus cannot replace other activity; it can only supplement it. It can help organisation, especially with a large class, because it enables the teacher to give a practical task to a small group while she is busy with some of the others. She must keep contact with this group, returning to it from time to time to talk about what the children are doing and to help their learning forward. But once she has trained them in the use of this material, she can treat the activity as a kind of directed play, in which the children are engaging in the particular aspect of learning that she has in mind at the time.

One other purpose of card apparatus is that it can give children practice that helps to consolidate learning which has arisen from practical activity, and practice is necessary even at this early stage.

Card apparatus cannot yet be used as a source of information in a project, as a means of directing children towards sources of information, or for asking questions that are presented in words.

### 3. *The Reading Difficulty*

This problem presents itself in a particularly acute form in the very early stage when few, if any, of the children can read at all; but even when they have begun to read the problem continues for some time, because it often takes so many words to explain even a comparatively simple mathematical activity. There is no way in which the reading difficulty can be entirely overcome, and the teacher or the helper will have to reckon on interpreting for the children any words that are on a card. The problem eases if the class is vertically grouped, because there are likely to be some competent readers to help those who are only just beginning; but this kind of help is probably not available if single-age grouping is in operation.

However, although the reading difficulty cannot be totally avoided there are some steps the teacher can take to minimise it.

(i) She can have as few words as possible on a card. For example, if there are some apples to be counted, 'how many' is better than 'how many apples can you see'. It could be argued that as the child cannot even read 'how many', there is no point in putting any words on the card at all. Just have the apples and tell the child to count them.

However, we would defend the use of *some* writing, because it is part of the process of beginning to familiarise children with the outline and meaning of common mathematical words and of associating the spoken word with its graphical form. Learning to read should not be confined to 'reading apparatus'. Too many words on a card, however, tend to be counter-productive.

(ii) The careful use of colour can help the child to learn what the card asks him to do, even though he cannot actually read the instruction. Some instances of how colour may be used in this way are given in the examples of card apparatus in the next few pages.

(iii) Whenever possible, words should be illustrated. The ability to draw is wonderfully helpful to the infant teacher, but if she is unhappily deprived of this talent there is a good deal she can do with geometric shapes, with pictures cut from books and magazines and with the packs of pictures available from educational suppliers. Illustrating takes longer this way, but it serves its purpose – and is essential.

*4. Variety*

One of the important theoretical requirements that emerged in earlier chapters is that if the child is to abstract mathematical principles from his experience, he needs to encounter the principles in a variety of unlike situations.[1] This applies as much to the use of card apparatus as it does to other activity. Therefore in designing card material the teacher must try to present the same principle in many ways which, to the child, are different. The game is different, the pictures and the approach are different, the arrangement on the card is different – but the underlying principle is the same. The reader will readily discern this, in some of the examples given below.

Variety is not only mathematically necessary; it is also essential if interest is to be sustained and boredom avoided while the principle is being learned.

5. *Keeping Card Apparatus in Order*

Because of the content of mathematics at the beginning stage (e.g. sorting, classifying, ordering, matching) some of the card apparatus that children use consists of several cards which together make a set. These can easily become separated, muddled and even lost. The easier it is for children to keep it in order, the more value they will gain from it. It becomes very confusing if some of the bits go astray or if parts of another piece of apparatus turn up with a card to which they do not belong. One or two tips may help the teacher to minimise the problem.

(i) Punch a hole in the corner of each card and thread together all those belonging to one piece of apparatus with a treasury tag (i.e. a short lace with a bar at each end). Show the children how to unthread the cards, and train them to re-thread them when they have finished.
(ii) Identify all the cards belonging to one set by marking the back of each with a symbol, such as a green cross or a yellow ring. Train the children to sort them out at the end of the activity, before re-threading them.
(iii) Make a check from time to time to see that a piece of apparatus is still complete, and replace at once any bits that are missing. This is especially important with any kind of self-correcting apparatus, because its purpose is destroyed if it is not all there.

*Examples of Suitable Card Apparatus*

(Cards with words on them are, where necessary, identified in some way, as suggested in the first two examples.)

1.

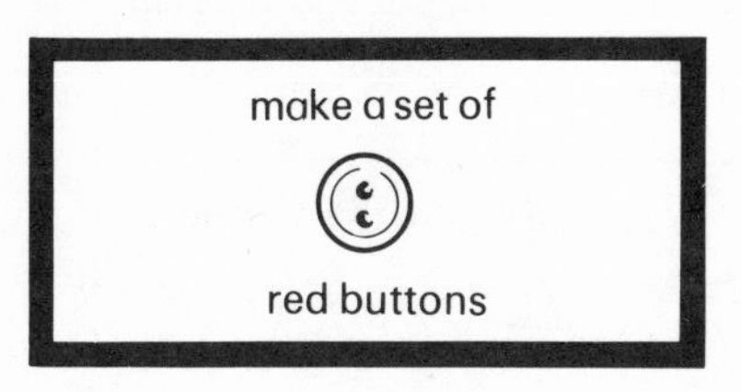


These cards might all have blue borders. The child learns that the blue-bordered cards are those which ask him to 'make a set of'; and the illustration tells him what to put in the set.

2.

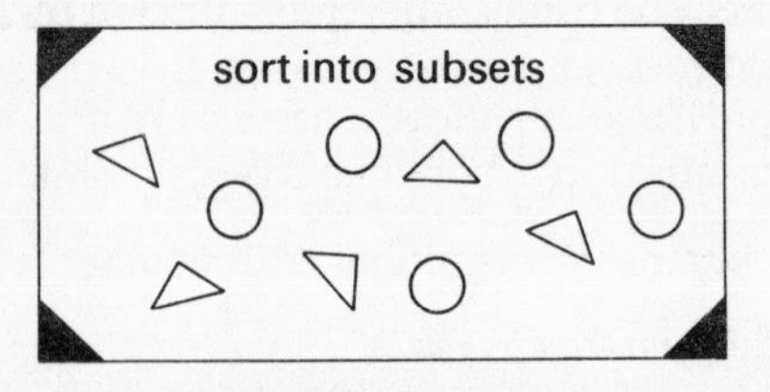


Green corners mean 'sort into subsets'.

*Variations of sorting into subsets*

(i) The card in the illustration would show circles and triangles all of the same colour; they would be sorted into subsets with shape as the criterion.

(ii) Having sorted into subsets, the child joins one-to-one with pieces of string to compare the number of elements in each set.

(iii) Another card shows triangles of more than one colour, including some that are red; and other shapes, all of which are red. The child sorts, using the criteria of triangles and red. Some triangles are red, so there is an intersection of the common property of red triangles.

3. Sets of pictures which have a natural connection, for matching one-to-one. (See page 58, para. 2.)

4.

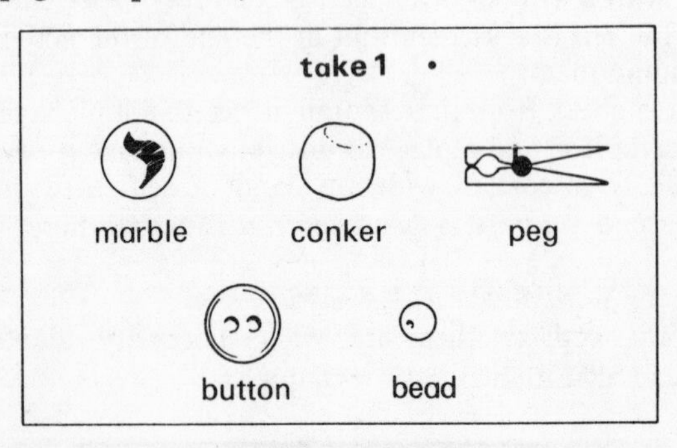


(For use with activity on page 58, para. 3).

5.

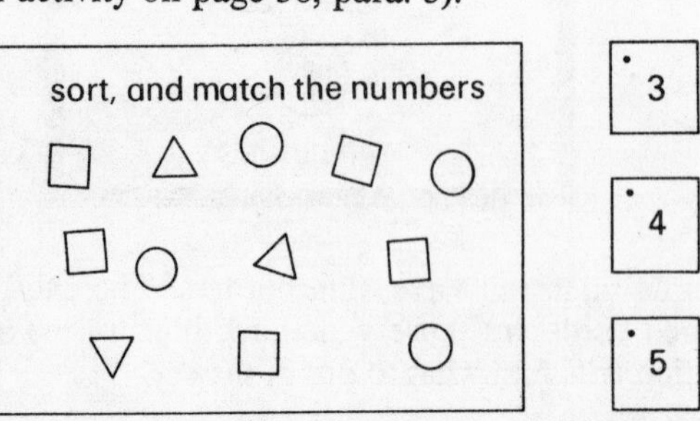

When sorting and matching are completed, the child draws one or two of the sets and writes the appropriate numeral.

6. The farmyard game is a versatile game which, though not competitive, can be used at different levels for sorting, matching, one-to-one correspondence, number recognition and number value (also word matching).

A second card shows 2, 3 and 4, and a third 3, 4 and 5

On separate small cards are the requisite number of hens, pigs and ducks, the numerals and the words.

Together the cards make a game for three children. The small cards are in a tray in the middle of the table. As each child completes his sorting and matching, he draws one of his sets of pictures and writes the appropriate numeral.

7. A set of number jigsaws: for the child whose counting is still uncertain, the jigsaw element provides a clue for matching the correct numerals to the objects. He also matches a counter to each picture to help one-to-one correspondence, and he records by drawing the one he likes best.

8. A set of five cards, each containing pictures of from one to five objects, or three to seven, or whatever numbers are suitable for the child's level. Five small cards show the numerals. The child matches a counter to each picture and a numeral to each card; and then draws the one he likes best.

9. Fishing games.

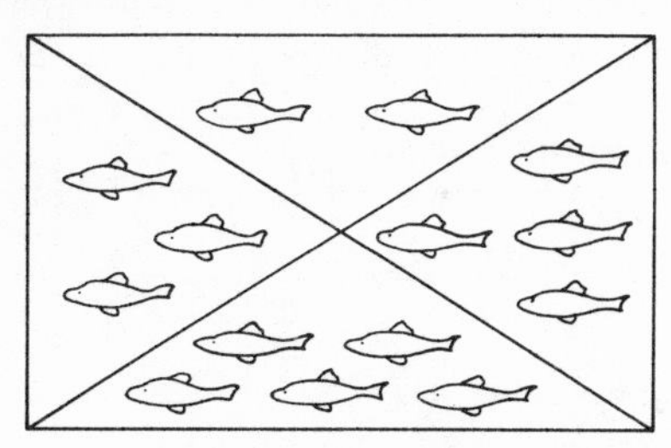

Level 2 As Level 1, but with the addition of separate small cards bearing the numerals, to match to the number of fish in each section.

Level 3

4 5 6 7

There are fishing rods with magnets on the end of the lines and a 'fish tank' containing cardboard fish with staples in their noses. No numerals appear on the fish. At Level 1, when the objectives are to help matching and one-to-one correspondence, the child matches fish to fish.

Level 2 begins to give experience of number value and number recognition, and this is taken further at Level 3, when there are no fish already drawn on the card to provide the extra clue.

*Note:* Children are sometimes given fishing games in which each fish bears a numeral, and the children 'go fishing' as a mathematical exercise. However, this is not of mathematical value. The children can fish quite happily without learning anything about the numerals. The mathematical objective of the game should be clear to the teacher, or her time in preparing it will be wasted.

10. A game of dominoes, played by two children.

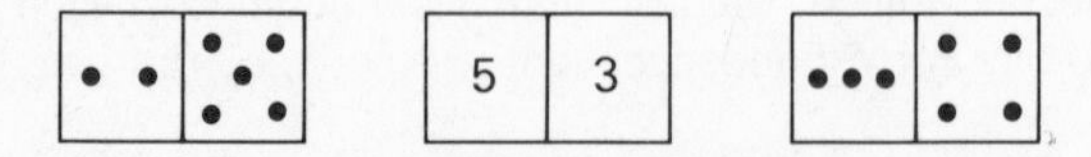


If the teacher's artistic skill will rise to it, pictures are more interesting than dots. An alternative is to use small geometric shapes.

11. Cards are useful for giving children additional practice in number recognition and number value. As an alternative to the activity described on page 86, para 7, pieces of kitchen paper cut to size can be stapled on to cards and torn off as they are completed.

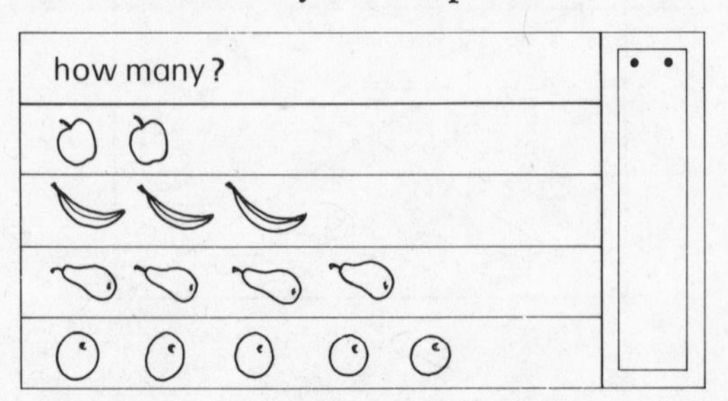

The converse activity is to write the numerals on the card, and the child draws the required number of objects on the paper.

12. For matching shapes:

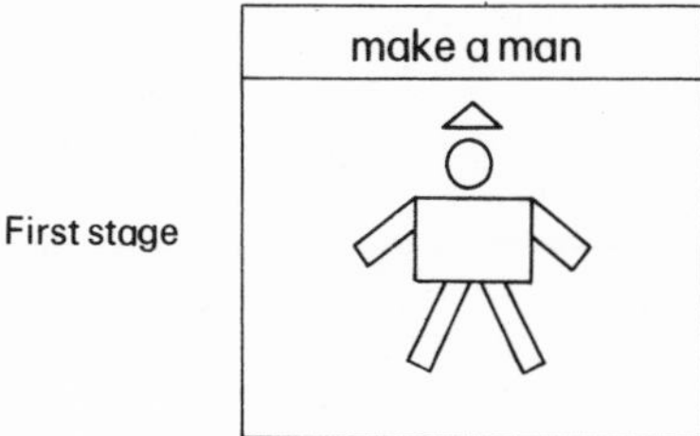


At a later stage the child is asked to write down the names of the shapes he used, and how many of each.

Later still, he is asked to see if he can make the picture with alternative shapes (for example, he may find that he can use two triangles instead of a square, or two or more squares instead of a rectangle).

13. The child is given a lump of plasticine and a card.

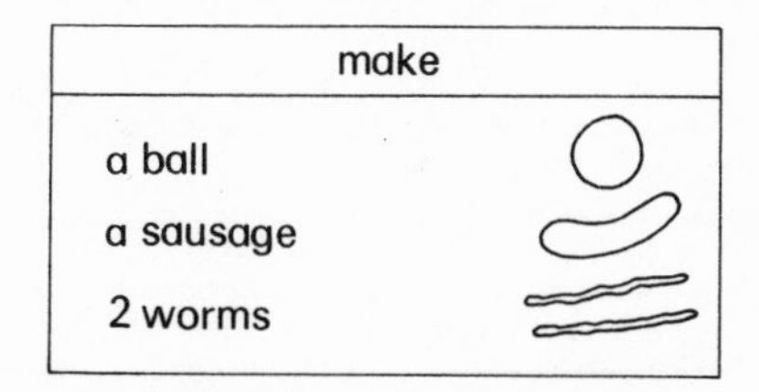


He will have to be told to use the same lump of plasticine each time, and a few minutes discussion with him is necessary as he does so. This will lead him towards an understanding of the conservation of substance, which will come later on.

## NOTE

1 Dienes, *et al.*

*Chapter 11*

# The Changing Scene

## THE CHILDREN

The understanding of the conservation of number is a very important milestone in the child's mathematical progress, because until this concept is firmly grasped no operations with numbers are possible; and operation with numbers are fundamental to the task of building on the foundation skills. However, although the scene changes when we move into this next stage, it is not, of course, an immediate metamorphosis in regard to every child in the class.

Most of these children are still very young and inexperienced. Indeed, as has already been suggested, it is not rare to find those who understand conservation with regard to number before they start school. Furthermore, some will be so near to it that they may need only a few weeks of pre-number activity before they can do so. On the other hand, there will be children for whom a great deal of foundation experience is necessary; and by the time they reach this mathematical stage they will be older, possibly a little more advanced in other ways, and perhaps in need of much reassurance from the teacher to counteract feelings of inadequacy and some loss of confidence when they become aware that their fellows are going ahead more rapidly. It is important for the emotional well-being of these children that the teacher should try to devise some pre-number activities that at least *appear* to be different from those which the more advanced children have already left behind, to give them such extra help as the many demands on her time will allow, and to make opportunities for drawing attention to other ways in which their skills are more apparent.

However, for our present purposes of building on the foundation skills, we will concern ourselves with those children who have developed an understanding of the conservation of number; and we will take them to a stage which many of them will not reach until they are well on their way towards the junior school. We will, however, stop short of the mathematical learning that is sometimes possible for the more advanced children by the end of the infant school period. Learning at this level is already much more fully covered in books on the teaching of mathematics than is that which applies to children at the earlier stages.

Despite the fact that the concept of the conservation of number

physical and intellectual reasons, all of which are important for their natural development and their personal fulfilment.

2. They need the freedom to investigate and experiment, to find things out by trial and error, to learn to anticipate the results of their actions. In play of this kind mathematical learning *is* one of the objectives. It may seem 'free' to the child; play, after all, is a natural activity in which he likes to engage. But if it is to advance his mathematical learning, the teacher must devise play situations with this end in view; and she must try to precipitate learning by suggestion, by asking open-ended questions from time to time, by leading the child towards a line of investigation that will be mathematically productive.

If she is not clearly aware of the purposes of play, the teacher may find that it tends to become totally aimless and she would have difficulty in justifying it on educational grounds. It is especially important that she should guard against this as children's learning advances, because aimless play can become very disruptive and annoying to others in the class.

Discussion continues to play a prominent part in learning. Its function as a vehicle of recording diminishes as children become more competent at recording in other ways, but as a means of verbalising experience, of developing thinking processes and of 'rubbing minds' with the teacher and with other children, its influence is strengthened, if anything, as time goes on. Therefore discussion can never be replaced by graphical recording, though this is not an argument against the necessary development of graphical recording which is itself important as a learning medium.

Though we regard the conservation of number as a notable landmark in the child's progress, we recognise that he has not yet grasped the principle of conservation in other aspects of mathematics. He will do so, in some of these aspects, during the period now under discussion; and an understanding of the conservation of number will help. For example, the child can now, given the experience, measure length with a unit of some kind and find that the table is eight units long. He could not make mathematical sense of this when he did not know what eight was, but his knowledge of number now makes possible for him certain activities which will lead him to an understanding of the conservation of length. Indeed, a great deal of learning will take place, in a comparatively short time, because the child is now equipped with one of the basic tools that he needs at this mathematical level.

## THE LEARNING ENVIRONMENT

As the children become more capable of accepting a degree of responsibility for their own activities, they will move out of the classroom or the more intimate learning area for some of their mathematical experience. At first this may be no further than the corridor, where there is more space to keep the shop. But later on, as we shall see, many valuable maths activities can be pursued further afield, particularly outside if weather and the geography of the school grounds permit. However, careful preparation and effective training of the children are essential if these pursuits are to advance any knowledge of mathematics. A trundle wheel is a useful instrument of measurement; but you cannot have half the class taking off for the playground at the drop of a hat, because trundle wheels are easily transformed into rockets or tanks; some learning might take place, but it would not be especially mathematical! The inexperienced teacher would be wise to defer this kind of activity until she feels confident in her training of the children.

The classroom area will undergo change as the children's learning moves forward. Certain additional items of equipment will come into use. Some of these are fairly small – rulers, weights, clock faces, etc. – which are liable to disappear amid the welter of material that finds its way into a busy classroom. However, if the children are to be successful in doing progressively more for themselves, they will be very dependent upon the orderly arrangement of the room. Searching for a ruler is a waste of time, and it is a fine excuse not to search very hard. If space permits, it is extremely helpful to have a maths bay or area where all or most of the equipment is kept. This has obvious organisational advantages; it is also educationally sound. Children engaged on mathematical activities find themselves together and there is a natural opportunity for the exchange of ideas or a discussion of intentions. Certainly this does not invariably happen, and it can happen in any other part of the classroom whether or not there is a maths bay. But opportunity helps, and the more of it there is the more likely it is to be used.

So if there is space for a maths bay, the teacher should organise one at this stage. The children will know where to find the materials they need, the assignment cards, the wall charts that help with the vocabulary of maths, the books which the teacher has provided. If space is a severe limitation, the impossible cannot be achieved and, as always in difficult conditions, the teacher must do the best she can. If there is no room to set the equipment out on work tops, it must be stored in boxes which are clearly labelled (and until the children can read, words alone are not enough; the box containing coins, for example, must have a coin pasted on the outside). There is no doubt that the more accessible the materials, the more likely the children are to use it and the easier it is to insist that they fetch and carry for themselves.

## TEACHING PROCEDURES

As a general pattern, we suggest in the following chapters the introduc-

# Chapter 12

# Number

Before looking at the mathematics open to children now that they have an understanding of the conservation of number, we should first say a little more about the practical implications of what conservation means at this stage.

Children may understand conservation to about five or so before they can do so with larger numbers. So the teacher has a choice of three courses of action:

1. She may continue with pre-number work, deferring the introduction of number operations until the children can understand conservation to about ten. She may encounter the difficulty that some of them are beginning to tire of pre-number experience and want to apply their advancing mathematical understanding to different processes that will make use of it. So she must use a certain ingenuity in devising some new pre-number activities in order to hold the children's interest.
2. She may introduce number operations to about five, continuing with pre-number work only for the larger numbers. This goes some way towards overcoming the difficulty suggested in the first alternative, but she must guard against the feeling that the specifically pre-number activities with the larger numbers are 'baby-stuff' which the children want to leave behind. Some of these budding mathematicians can be rather sensitive about this!
3. She may introduce number operations to five, and allow for the extension of conservation to come about as a result of general mathematical activity. Provided she is careful to restrict number *operations* to the smaller numbers, ensures that other activity is unobtrusively giving the children adequate experience of the larger ones, and treats them individually so as to take account of such factors as temperament and confidence, this may be the best solution. The teacher's knowledge of the children she is teaching will determine the right course of action for a particular child.

### ADDITION

This is the number operation with which children usually begin, though an element of subtraction enters into some of the activities – thus preparing children for the next stage of progression. There are many ways of presenting the operation of addition, but an example is suggested that has been found to work quite well in practice.

The children in the group are each given a number of cubes – let us say five. They are asked to arrange them in as many different ways as they can. With some experience of this activity they find that they may arrange their cubes as two and three, or three and two, or one and four, or four and one. Quite apart from consolidating their understanding of conservation, they are also coming to understand the commutative law of addition in a natural and painless way.[1]

The children are encouraged to experiment with this activity, using different numbers and other materials besides cubes. At this stage the teacher is giving the group a good deal of attention, with discussion and questions; but she leaves them from time to time to explore the possibilities without her direct help. After an experimental and purely practical period, which may not last very long, the teacher shows the children a way of recording this operation. She also makes it possible for them to collect the activity for themselves and engage in it without her having to set it up for them each time. One suggested method is as follows:

1. In the maths bay, have a box or a wall pocket containing some 'work cards', e.g.

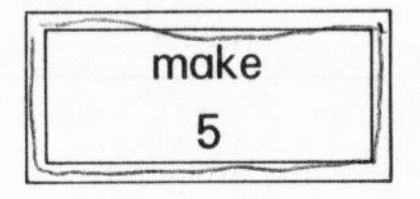

A second set, on different coloured card, can show the numbers from six to ten for children to move on to when they understand conservation with these numbers.

2. Adjacent to these cards, have a box containing small cards, about 3 cm square, depicting the required number of numerals from one to five (or one to ten for the later stage); and a box of cubes and other suitable material.

As the child's understanding extends, so the numbers on the cards will become larger.

*Other Forms of Presentation and Recording*

In some of these forms, a great deal of the teacher's time can be saved if she has access to a duplicating machine. If this is not available, she must either put the material on cards which the children copy before completing the operation, or she must put it individually into their books (which is very time-consuming).

1.

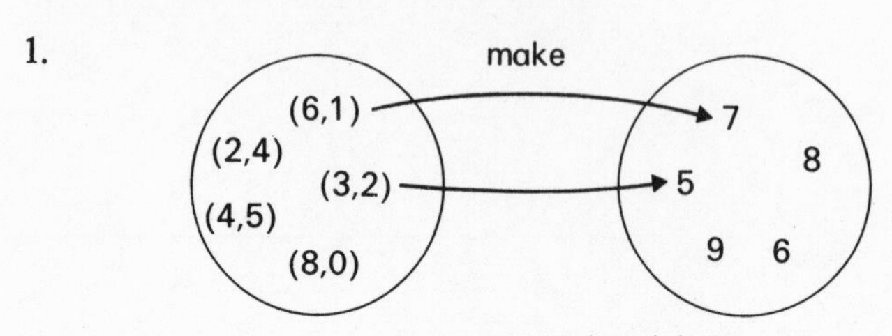


or it might be set out as:

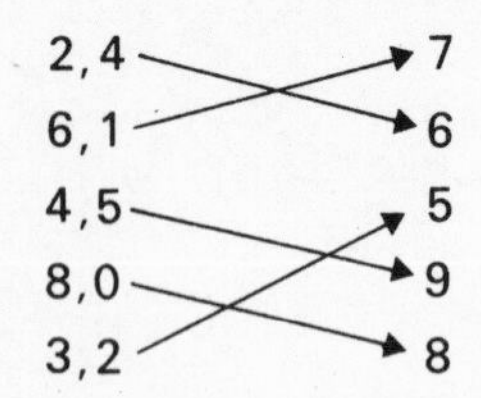


With arrows, the child joins those on the left to the total on the right.

Alternatively, the circle on the right is left blank, and the child inserts the total at the end of his arrow.

2.

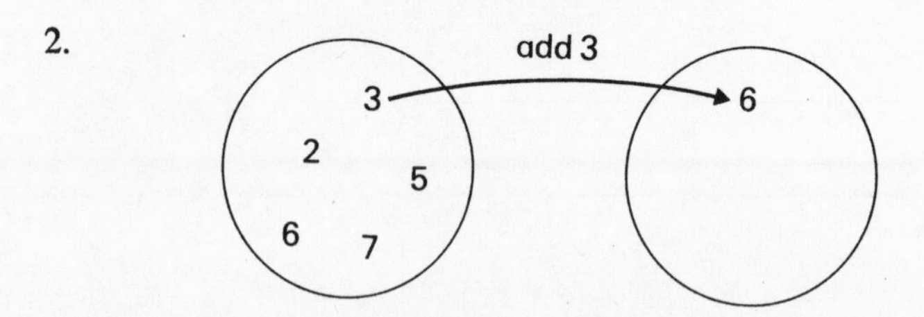


Again this may be presented in either form as in 1.

3. Coloured, self-adhesive dots are available quite cheaply from stationers and educational suppliers. They provide an interesting variation for recording, which children enjoy. The child sticks the required number of dots into each part of the set and then gives the total.

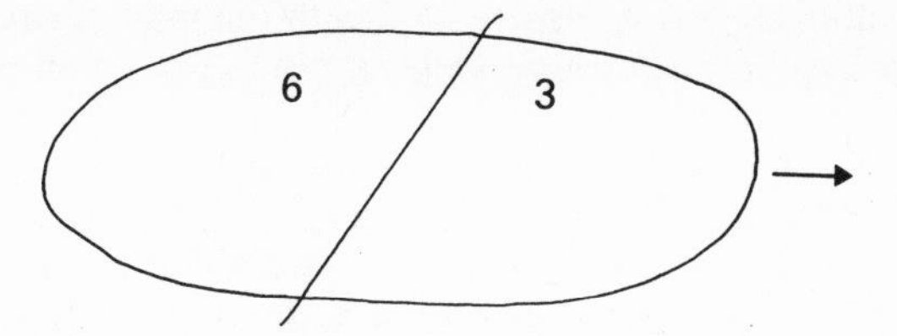


4. The teacher sticks a given number of dots; quite widely spaced, into a set. With his pencil or a short piece of string the child partitions the set in as many different ways as he can, recording each combination that he finds.

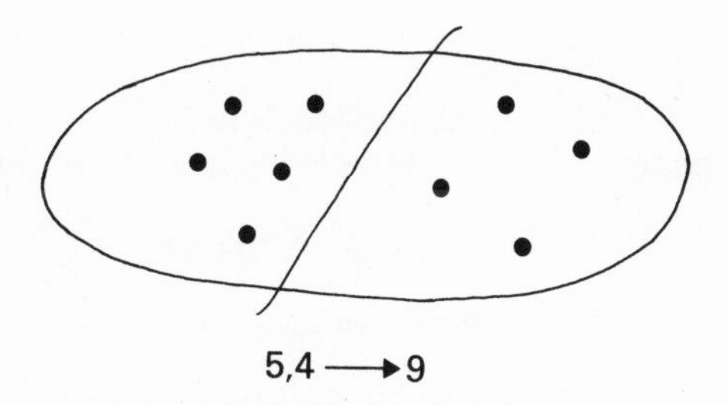


5. Introducing the operation of subtraction.

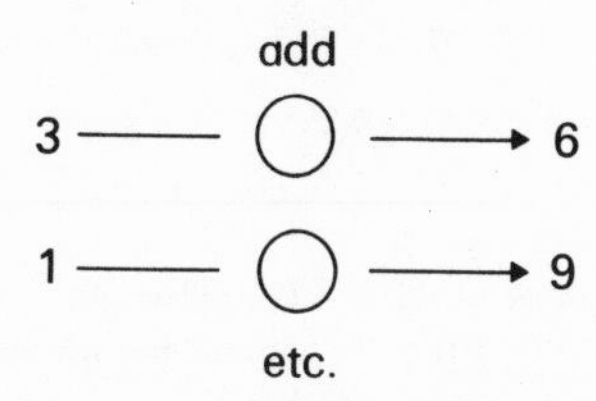


6. Counting on.

Count on

3 —— 4 ——→
6 —— 3 ——→
5 —— 0 ——→
7 —— 2 ——→
4 —— 6 ——→

Using a number strip, the child counts the first number, counts on the second, and records the sum with the total. Block graphs can be used very effectively for counting on; so can interlocking cubes and segmented rods.

7. A number strip may also be used with the kind of small cards suggested on page 115 (but here they have no connection with tens and units).

8. Cuisenaire and other similar rods, which can now be given numerical associations, are excellent for assembling the component parts of a number. When the children are sufficiently skilled, they can be taught how to record the various assemblies on squared paper, using coloured crayons to make representations of the rods and attaching a numeral to each representation. (For Cuisenaire rods, 1 cm squares are required.)

Children can also make 'staircases' with the rods from one to ten, then add one unit to each stair, two units, etc. – and see what happens to the numbers they are building.

Two or more different rods can be grouped end to end on squared paper, the squares coloured in, and the number of squares of each colour added.

*Some Addition Games*[2]

*1. Spinning Tops.* These may be bought, or may be cut from fairly substantial card.

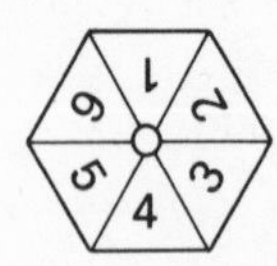


A pencil that is too small for further use is put through the hole in the centre. The child spins the top twice (later on three times), and adds the numbers at which the top comes to rest.

*2. Dice.* The child throws two dice, and adds the numbers that fall uppermost. Later on he totals the numbers on three dice.

*3. Fishing.* The fish now have numerals on them, and the child adds those on two or three fish that he retrieves.

*4. Scoring games.* These give practice in addition as the children's skill increases and they can add more, and larger, numbers. Examples of such games are tunnel ball, skittles, a board with numbered hooks on to which the children throw small rubber rings; the fishing game can also be used this way.

There is no reason why children should not play these games while they are still at the stage when they need counters to help them with adding. As they score *3*, they take three counters; and they total the score by adding up their counters.

Children should move on to addition to twenty when they are ready for it, using their apparatus for the calculations for as long as they need them. One or two children may go beyond twenty, even by the stage of middle infants, and they should be introduced to the idea of tens and units. So it is perhaps worth taking addition a little further, recognising that this will not apply to the great majority of children before the top of the infant school.

The first step is for the child to be presented with material that helps him to recognise that if 2 and 3 are 5, 12 and 3 are 15, 22 and 3 are 25, etc. The best way for him to do this is by using a number strip. This should be designed as follows:

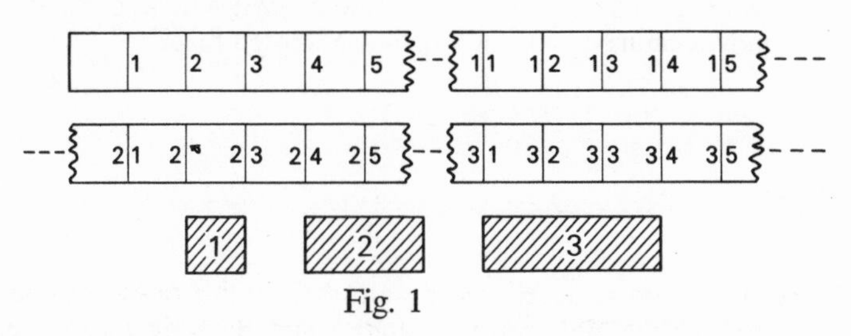


Fig. 1

The small cards correspond with the length of the units represented on the number strip.

By placing the appropriate small card on the number strip, the child can see what happens, e.g.

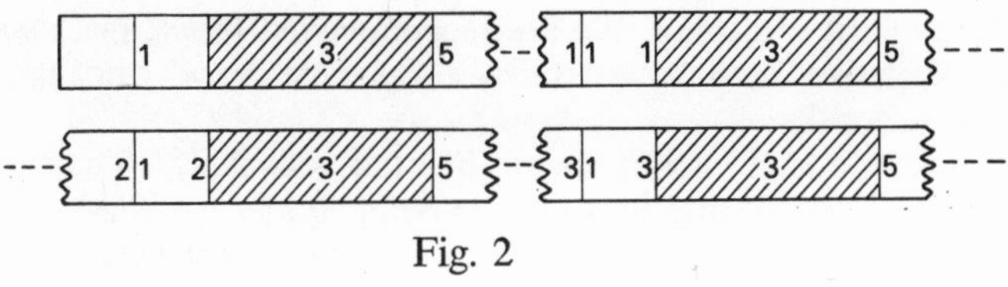


Fig. 2

The numerals on the number strip are arranged in such a way that when the child positions the '3' card on the strip, as shown in Fig. 2, the numerals that are revealed on either side of the card give the 'answer'. This enables him to see the pattern of 12+3, 22+3, 32+3, and so on.

An understanding of this pattern is necessary before children can move on to the next step, which is the addition of tens and units. For this purpose it is helpful at first if their apparatus now consists of small sticks or rods which are easily put together in bundles of ten with rubber bands. The child should first count, say, fifteen sticks, put ten of them in a bundle and see how many are left over. The 'leftovers' are called *units.* So he has one bundle of ten and five units, which is why he writes the number as 15 – one five. He should now engage in the same activity, using counters which sit in piles of ten, and leave odd units; ten pennies which can be exchanged for a 10p coin, the units remaining in pennies; ten cubes which can be locked together, ten shells which can be put in a bag, ten unit rods which can be exchanged for a *ten* rod.

When the child is at ease with making a bundle or collection of ten, finding the leftover units, and writing the numerical symbol, he is given two similar numbers to put together – say twelve and thirteen. He counts the twelve sticks, puts ten in a bundle, and finds that he has two units left over. He then counts the thirteen sticks, and with one bundle of ten he has three units left. Putting the two lots together, he now has two ten-bundles, and five leftover units, and he writes them as two five – 25.

He moves on next to sums which are written thus:

$$\begin{array}{r} 12 \\ +13 \\ \hline \\ \hline \end{array}$$

No carrying is required yet, because really this does not come until later; but if the child learns to understand the *principle* of tens and units, carrying will, at a later stage, make much more sense to him.

*Plus and Equals Signs*

One detects a belief that these and other similar signs no longer have any place in the infant school at all. This is one of those generalisations which is hard to defend. We are recommended to use just as abstruse signs, such as $<$ and $>$, at a fairly early stage. So why not the faithful old $+$ and $=$?

The reason is, no doubt, the problem of vocabulary. The very words *plus* and *equals* are not part of the young child's everyday vocabulary, whereas *bigger than, more than, less than* are at least familiar and are likely to acquire mathematical meaning sooner than one can expect with unfamiliar words. (However, we ask ourselves whether the word *add*, until it is introduced and explained, has any familiarity either, and yet it appears regularly in suggested ways of presenting the operation of addition.)

There is a strong case for arguing that the traditional signs *should* be presented, as alternative mathematical statements, once the children have enough experience for a different statement not to confuse them. The difficulty in the past was that the $+$ suddenly appeared when sums began, often without any real explanation of what it meant beyond reference to the fact that it was *adding* (which did not mean much to children either). If, however, the teacher explains that the + sign is simply a quick alternative to the words *add* or *and another*, or to the arrow which has been used to denote them, we see no reason why the child should not readily absorb it into his repertoire of graphical representation. He is going to meet these signs sooner or later, because they are an accepted part of mathematical language. They are quick to write, and when the child moves on to the other three rules they spare him some (but not all) of the wordy rigmarole of *take away*, *lots of*, *multiply by*, *share between*, etc. Certainly these words have a use in the initial introduction of the processes, but in the long term they are clumsy alternatives to the traditional shorthand symbols.

The *equals* sign is a little more troublesome. Whereas + always means *add*, = sometimes means *makes* and at other times it may, to the child, mean *leaves*. These are two contradictory linguistic statements. It is therefore suggested that the + is introduced fairly soon, but that the use of = is deferred until the child can accommodate its ambiguity without confusion. Thus an alternative form in the presentation of the process of addition would be $2+2\rightarrow4$, the meaning of the + having first been adequately explained. It can also be used sometimes in arrow diagrams to explain the arrow, as an alternative to the word *add*.

### SUBTRACTION

Children will have begun subtraction long before they reach the level of competence in addition which we have just been describing; so we will now return to the point at which subtraction may first be introduced in a systematic way.

Although there were elements of subtraction in some of the earlier addition activities, the child will not have thought of them as anything other than addition. As Williams and Shuard point out, children do not at first relate subtraction to addition. They see them as quite separate and unrelated processes, which is understandable. With addition, after all, the items being added are there to be counted; with subtraction they are not. Those which are taken away have gone. To many children the connection between the two must, to say the least, be tenuous.

'Sweets which have been eaten have gone for ever. But if a child is to deal mathematically with subtraction situations, this process must become reversible. . . . He must . . . be able to separate his 5 sweets into 3 to be eaten and 2 to be kept, so that he can see that he is dealing with another example of addition. . . . Subtraction is only fully understood when it is seen as an aspect of addition.'[3]

There are several ways in which we can help children to gain experience of the relationship between these two mathematical processes, and we should present them with this experience before asking them to do any kind of subtraction 'sums'. The underlying principle of such experience is that the child should start with the *whole*, and then set aside some of the *parts* in such a way that he can see they still exist. This kind of experience (together with teaching in the sense in which we have been using the word) would enable him to abstract the connection between subtraction and addition. In recording this abstraction, the use of the shorthand + and − signs are extremely helpful in avoiding all the written words that are otherwise necessary. For many children at this stage, reading and writing skills are still very limited and at best slow and laborious. If they have already met the +, the − should present no great difficulty provided, once again, its meaning is clearly explained. There seems so little justification for adding to children's problems in one way, in order to avoid them in another.

It is hoped that the child will come to see the connection between the following four equations:

$$3 + 4 \rightarrow 7,\ 7-3 \rightarrow 4,\ 4+3 \rightarrow 7,\ 7-4 \rightarrow 3.$$

In view of this, there is some difference of opinion as to whether addition and subtraction should be introduced simultaneously from the beginning of number operations, or whether subtraction should follow when the operation of addition is understood. There can be no hard and fast rule about this, and again only the teacher's judgement can guide her as to which method she finds most successful in practice. There is sometimes a tendency for children to become confused when two processes, even though they are related, are presented to them at the same time; but if the teacher can overcome this difficulty there is no reason why she should not introduce the two operations together if she prefers to do so.

It is perhaps worth noting, though, that one could argue similarly about addition and multiplication. Logically, these should surely be presented simultaneously, since multiplication is a form of addition; but mathematically logical or not, there is a limit to the number of mathematical ideas that a child may realistically be expected to absorb at more or less the same time.

However, a preference for introducing addition first does not invalidate the argument that subtraction, if the child is really to understand it, must be seen by him as an aspect of addition; and that he must recognise the principle of reversibility which connects the two. It is therefore logical to begin by providing him with the experience that will help him to make the connection. The following suggestions indicate some of the ways in which this might be done.

1. A similar method may be used (with obvious variations in recording) to that described on page 109 for the introduction of addition. This time when the child re-arranges cubes and other material, he is

asked to separate three of his cubes from the five, and see how many are left. The five cubes are *still there*, but they are separated into two groups. What happens when the two groups are put together again? Separate one cube, see how many are left, and observe what happens when they are put together. Try this with other numbers, and note the possible variations of separation and reunion. If the child is already familiar with addition, he will recognise that when he is reuniting the groups of cubes he is adding them; and he may, perhaps more readily, also recognise that he is merely reversing the process when he splits the main group by moving some aside, or by 'taking them away'.

2. Similar experience takes place in working with sets; partitioning a set in a variety of ways, and each time restoring it to its original number.

3. Rods and other structural apparatus offer another alternative. Which rods can be used to replace a blue rod? 'Add' them together again.

4. In shopping, six pennies are needed for a purchase. Of the ten in your hand, how many will you have left when you have paid for your purchase?

5. Many games can apply the principle of separation and reunion. Two children have a pack of ten cards. In turn, each child gives some cards to the other – how many has he given, how many are left? Similar possibilities in daily classroom life are so obvious that it is unnecessary to list them.

Children should not be asked to record these activities in writing every time, or they would be kept so busy writing things down that there would not be much time left for mathematical experience. (This, of course, applies to many other maths activities besides those concerned with learning subtraction.) However, we have already noted that recording is an important part of learning and we suggest ways in which this might now be done. The teacher's good sense will ensure that it is kept within reason.

If the child has already been dealing with simple addition, he will by now be familiar with some of the ways of recording early number operations. The small cards with numerals on them, which he had before, can be used again, but this time additional ones should be provided : some with a + sign and some with a −, or if the child has not yet been taught about these signs the words *add* and *take away*; there might also be some cards showing an arrow. Work cards would now take this form:

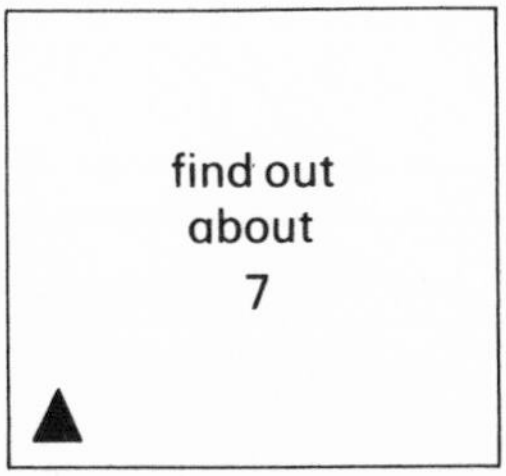

find out
about
7

▲

Since the child is unlikely to be reading yet, a small symbol on the corner of the card will identify it for him as a 'find out about' card.

He is taught to set out the activity in this way:

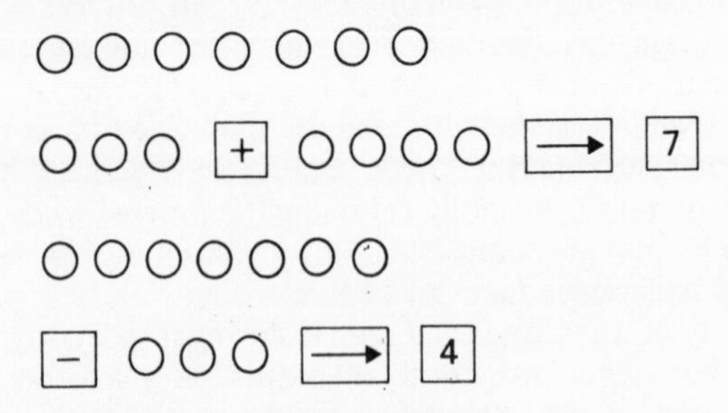


Finally, when written recording is required:

$3+4 \longrightarrow 7$

$7-3 \longrightarrow 4$

etc.

The child can be given subtraction assignments in many of the forms already described for addition, and some of the games will also be suitable. There will, however, be certain differences and modifications, e.g.

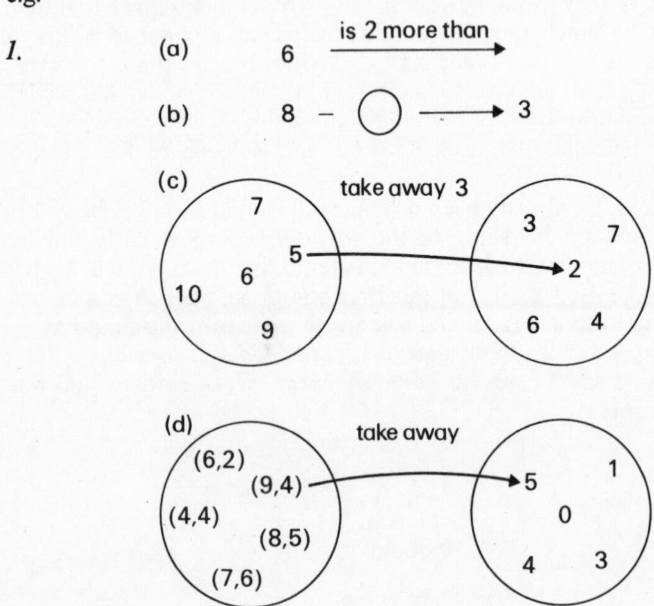

*2. Counting Back*
Counting back on a number strip.
Counting back on block graphs.
Counting back with rods.

*3. Flannel Board*
This can be made quite simply by gluing a piece of flannel or brushed nylon to some strong card, and providing a variety of cut-out pictures, also glued to card, and with a scrap of *Velcro* (about 1 cm square) stuck on the back of each. This provides variety by replacing counters in many of the activities already described.

*4. Rounds*
Cut two circles of card, one about 16 cm in diameter and one 18 cm. Cut out a 'window' from the smaller circle, and number the larger one as shown below:

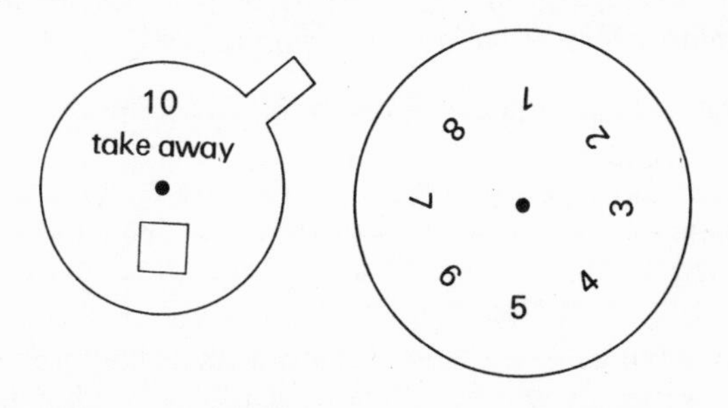


Fix the smaller circle on the larger one with a split pin. As the child turns the larger circle he records the sum indicated by the numeral which appears in the window, using counters to help him do so.
*Note*: This piece of apparatus can also be adapted for addition, multiplication and division.

*5.*

find the difference between

7 and 3
6 and 5
4 and 1
5 and 3

Apart from taking away, 'finding the difference between' is also an important aspect of subtraction. These are, however, difficult words for children to read, and the teacher really has no alternative but to give the card to the child and explain what she wants him to do – i.e. to make a heap of seven shells and a heap of three, and find the difference between the two heaps.

This is one of the cases in which the reading difficulty cannot satisfactorily be overcome, but the child should not be deprived of the experience of 'finding the difference between' because he cannot read.

*6. Games*

*Spinning Tops.* These can again be used, except that at first it is helpful if the child has two different-coloured tops, perhaps a blue one with the numerals 0–5, and a green one with 5–10. With the help of counters, he subtracts the numbers on the blue top from those on the green.

Later on, when the operation of subtraction is more securely established, he should use only one top, spinning it twice and finding out for himself which of the two numbers can be subtracted from the other.

*Fishing.* This can be adapted, as with spinning tops, by the initial use of colour to distinguish the fish, moving on to retrieving any two fish and determining how to subtract the numbers.

If subtraction is introduced after addition rather than simultaneously, the teacher should watch for the point when a child seems at ease with 'taking away' and then ensure that both processes are sometimes presented together. Addition and subtraction sums or activities should appear on the same assignment card, to reinforce the child's understanding of which is which. Activities which underline the principle of reversibility should also continue. After the initial stage, sums should be shown both vertically and horizontally, particularly when notation extends beyond ten, e.g.

$$14-3\rightarrow \quad , \text{and} \quad \begin{array}{r} 18 \\ -\ 6 \\ \hline \\ \hline \end{array}$$

Of those children who progress to tens and units in addition, some may also learn to apply the principle in subtraction. Take, for example, the sum

$$\begin{array}{r} 25 \\ -13 \\ \hline \\ \hline \end{array}$$

The child assembles two bundles of ten, and five units. From these he removes three units and one bundle of ten, leaving him with one ten and two units. In both addition and subtraction, the teacher may find that some children will quickly see that in sums of this kind it makes no difference whether they deal first with the units column or with the tens. This should not be frowned upon, provided the *principle* is grasped. We all do this for quick calculation in daily life, and when the child eventually moves on to the stage when a larger number of units must be subtracted from a smaller one (thereby necessitating the conversion into units of one of the tens) he will discover for himself that if he subtracts the tens first he will not be able to subtract the remaining units. (This assumes, of course, that he is given sufficient teaching and experience with concrete materials to enable him to recognise what the operation involves.) However, this touches upon a stage which is still some way ahead, but in the meantime the more advanced children should have the preparatory experience of subtracting tens and units in their simple form.

### MULTIPLICATION

Multiplication should be introduced as a form of repeated addition. For example, the child may first be given a variety of addition assignments such as 3+3+3+3→. These assignments can be presented in most of the practical ways already described. A number strip, for instance, or Cuisenaire rods or Unifix cubes illustrate clearly the principle of adding the same number several times.

We must now digress for a moment, for the teacher to consider the questions of terminology and the use of symbols. The mathematical argument is outlined in the *Schools Council Curriculum Bulletin No. 1*, which explains that the symbols +, −, × and ÷ signify particular mathematical operations; the symbol comes after the number which is *operated upon*, and is followed by the number which *performs the operation.*

'There is widespread confusion about the symbol "×". Its correct meaning of "multiply by" is very often replaced by "times" or "of", in which cases the number or measure which is being operated on follows and does not precede the symbol.

For example, 3×4 means 3 multiply by 4, . . . which is equivalent to 3+3+3+3 or "four threes" or "four times three". Also 4×3 means 4 multiplied by 3, which is equivalent to 4+4+4 or "three fours" or "three times four" '.[4]

If we accept this argument we cannot use the terms 4 *lots of* 3 or 4 *sets of* 3 and record them and 4×3, because the number which is being operated on follows and does not precede the symbol. On the other hand, words like *multiply by* are so remote from the everyday vocabulary of most young children, despite the argument that

'. . . if children have had sufficient experience with real situations to be ready to meet the symbols, then they are ready for the correct mathematical terms'.[5]

Certainly children should be taught the correct mathematical terms as soon as they are ready for them, but a linguistic difficulty does arise in this particular case. If the child sets out 4 groups of 3 acorns, his natural descriptive phrases to explain the situation before him are that he has 4 *lots of* 3, or 4 *sets of* 3. If we insist that he uses the words *multiply by*, we present him with a statement which, though mathematically correct, is not *explanatory* in his own terms. We should, of course, commit no mathematical crime if we were to use *lots of* or *sets of* without recording the operation by means of the symbol ×. It is the symbol itself which is incorrectly interpreted by words other than *multiply by* or *multiplied by*.

An alternative which can delay the introduction of × is to begin by recording multiplication as follows:

$$3+3+3+3 \rightarrow 12$$
$$4 \text{ sets of } 3 \rightarrow 12$$
$$4(3) \rightarrow 12$$

In due course the child would dispense with the first two of these statements, and use only the third.

He would be meeting the operation of multiplication in a variety of practical situations, including such activities as shopping. With this kind of experience behind him he should, in theory, arrive at the point when he is ready to meet the symbol × and attach to it the correct mathematical term.

However, it is surely open to doubt whether the child, even then, can really appreciate the mathematical difference between describing × as *multiply by*, or describing it as *lots of* or *times*. The words *multiply by* may continue for a long time to be merely descriptive of the symbol, rather than mathematically explanatory. The symbol is, nevertheless, a convenient form of shorthand in recording, which is a good reason for using it.

Whatever form of recording is adopted it is important to ensure that, from the beginning, the child's experience leads him to recognise the commutative property of multiplication, i.e. that 'four threes' are equivalent to 'three fours'.

'As with addition, once a child has discovered by making patterns with real objects (or using structural apparatus) that 4×3 and 3×4 are both 12, the number of different multiplication bonds which have to be remembered is almost halved.'[6]

This is more readily achieved if, in some of his assignments, the child is encouraged to focus on commutativity, e.g.

$5 \times 2 \longrightarrow$
$2 \times 5 \longrightarrow$
$4 \times 3 \longrightarrow$
$3 \times 4 \longrightarrow$
$3 \times 2 \longrightarrow$
$2 \times 3 \longrightarrow$

The child may do this with structural apparatus, with a number strip or track, an abacus, and by arranging objects in sets. When he goes shopping, he can be asked to buy five articles at 2p each and two articles at 5p each, and see how much he has spent on each group of articles. As a result of this type of experience the child may ultimately construct an *axiom* in the Dienes sense (see page 29, fifth stage) – in this case, an axiom about the commutative property of multiplication – which is then readily available to him for his mathematical purposes.

*Apparatus*
Much of the apparatus suggested for addition and subtraction is also suitable for multiplication. Some picture cards may be added at the initial stage, e.g.

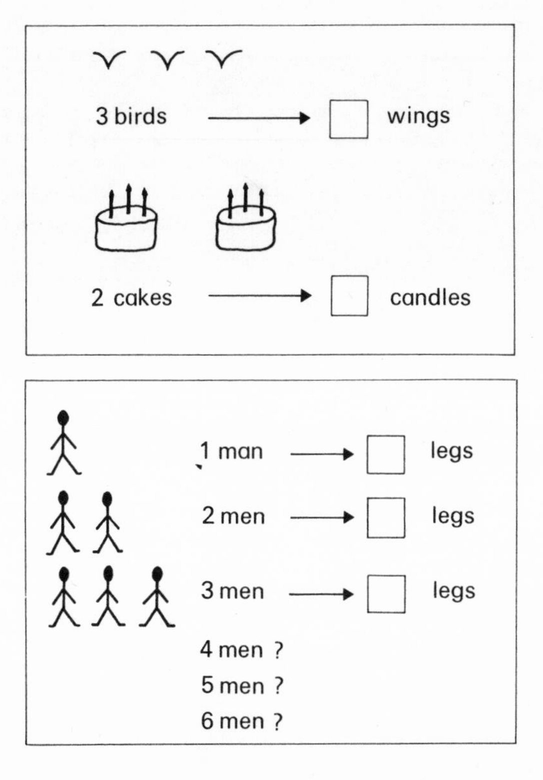

*Note*: Reading skills are now advancing, so words present less of a problem.

For use with a flannel board (see page 121):

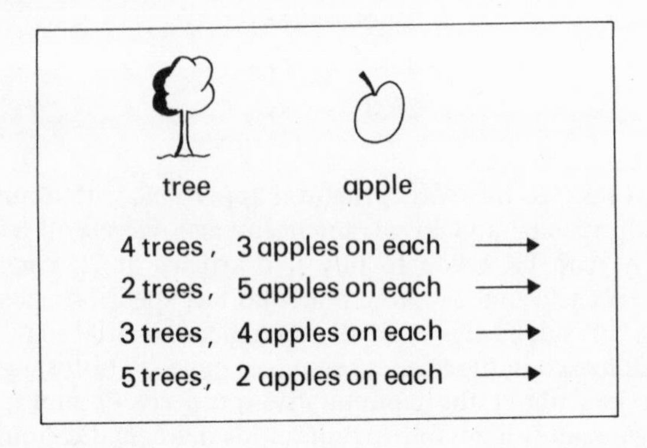


In time, children should be asked to think of as many ways as they can of making a number. For this purpose, structural apparatus is excellent. A 'make 6' card could be used as follows:

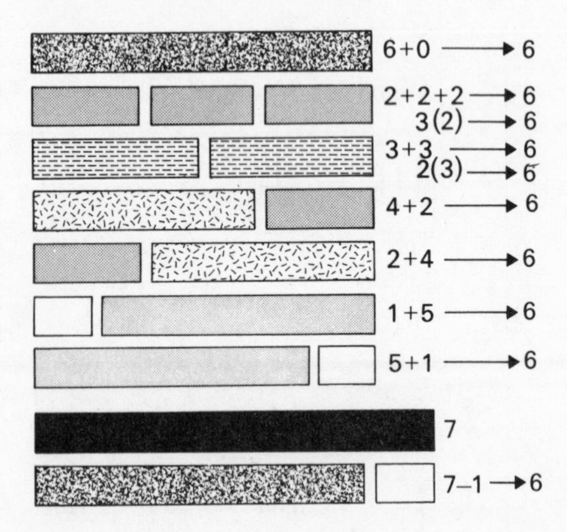


Care should be exercised in the use of number charts or tables for multiplication or addition.

| multiply | 1 | 2 | 3 | 4 | 5 |
|---|---|---|---|---|---|
| 1 | 1 | 2 | 3 | 4 | 5 |
| 2 | 2 | 4 | 6 | 8 | 10 |
| 3 | 3 | 6 | 9 | 12 | 15 |
| 4 | 4 | 8 | 12 | 16 | 20 |
| 5 | 5 | 10 | **15** | 20 | 25 |

$5 \times 3 \longrightarrow 15$

or $3 \times 5 \longrightarrow 15$

When, at a much later stage, children begin constructing tables, these charts help them to see how numbers are actually built up in tabular form. But they should *not* be used to replace counters or other concrete material for practice in computation. If they are, then all the child must do is learn the trick of finding where a vertical and a horizontal row meet, and he has the 'right answer' – without necessarily having any understanding of the mathematics of multiplication.

### DIVISION

Not many children are likely to do much about division in that part of the infant school with which this book is mainly concerned, so we propose to limit our observations to introducing the ideas of sharing and of dividing by grouping.

Sharing, for this purpose, means sharing equally; so attention must first be given to the meaning of the word *equally*.

If there should be, say, four children who are ready for this activity, the teacher may sit them in a group and give them twelve matchboxes which are to be shared equally between them. They may at first distribute the matchboxes in a haphazard fashion, and then find that each child does not have an *equal* share. They should then have the opportunity of trying to determine how to overcome this problem. Discussion will take place, and the teacher's questions will help the children to think about it without her actually telling them what to do. The activity continues with other numbers and other objects. The principle of *sharing equally between* begins to emerge.

The children may choose a number, like nine, which they wish to share between them. When they find that there is one left over, they begin to realise that not every natural number can be shared equally between a given number of children. This is a useful piece of learning.

When the four children have shared eight pennies, they find that they have four sets of two. So eight shared equally betwen four gives four sets of two, *because* four sets of two make eight. It is evident that for the full mathematical value of this activity to be extracted, discussion is absolutely essential.

The children will find it easier to record the activity of sharing equally if they are now directly taught to recognise, and to write, the words that are needed: *share, shared, equally, between, because.*

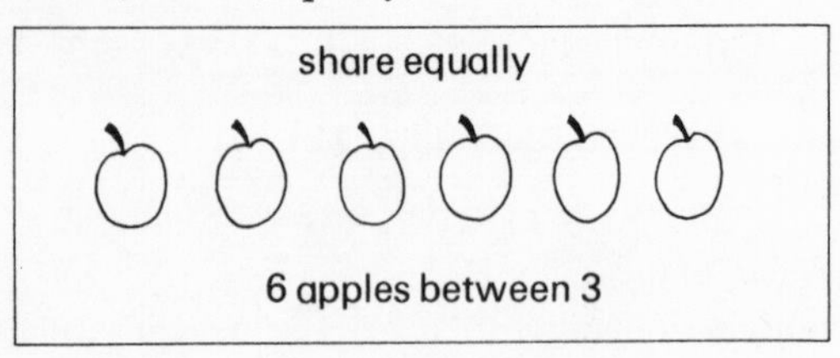


This may be recorded:

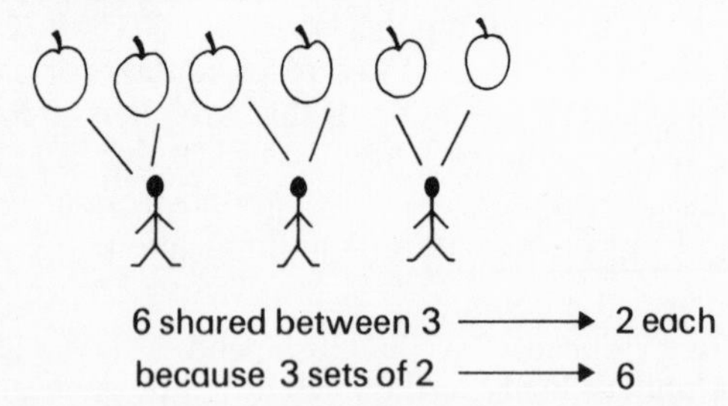


This may seem rather a clumsy way of recording division at the beginning stage. But if rote learning is to be avoided, and the mathematics of division properly understood, foundations such as these must be very securely laid. We recognise, however, that the writing part of the exercise may occupy a disproportionate amount of time, even though these children are by now likely to be much more skilled at writing: and this may detract from their enjoyment of the mathematics. The difficulty can be overcome if a duplicating machine is available, so that the child has duplicated sheets on which he can record, e.g.

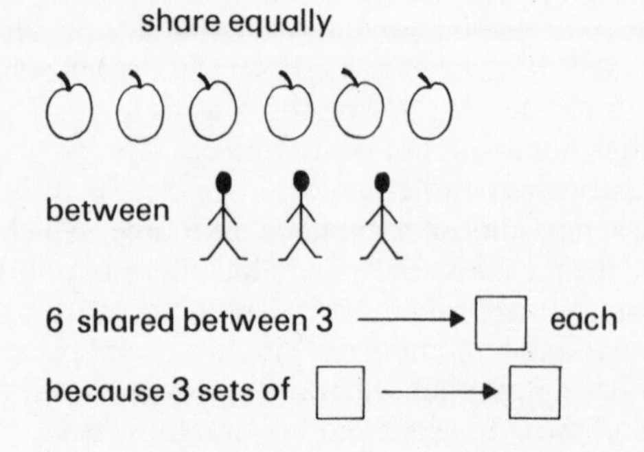

The child inserts the arrows between the apples and the people and also the numerals in the appropriate boxes. Several such exercises can appear on the same sheet.

A second set of sheets may be prepared thus:

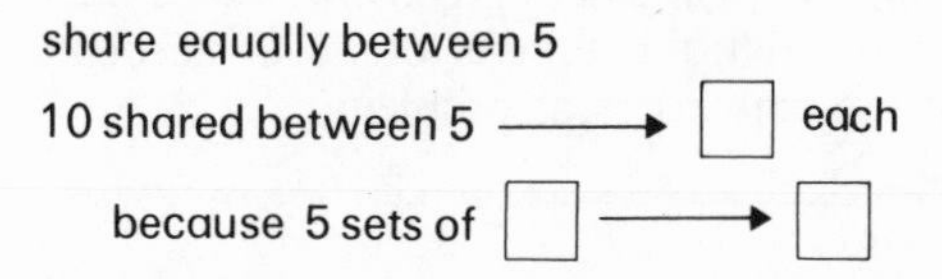


These sheets would be used for recording activities undertaken with a variety of materials (including their arrangement in sets and subsets). Separable structural apparatus such as *Unifix* is also suitable for division activities, and so is a flannel board.

*Sharing Boxes*

These are useful in the classroom at this stage. In a sharing box there are several small boxes, each containing a given number of objects. For example, the box labelled 'sharing box for 3' will have in its small boxes numbers of objects that are divisible by three – three conkers, six shells, nine beads, twelve coins, etc. Each small box is labelled according to its contents (e.g. '9 beads'), so that at the end of the activity they can all be restored and the sharing box made ready for use on another occasion.

Three children work together on the box for three, each child in turn selecting one of the small boxes and sharing its contents equally among the members of the group. Every child in the group records the result of the share-out.

When the teacher judges the children to be ready for a shorthand form of recording, she can show them that they may write $\frac{15}{3}$ instead of 15 shared between 3, the line representing the shorthand form of the words *shared between.* They can now record the statement that fifteen shared between three gives five to each member, because five threes are fifteen, as:

$$\frac{15}{3} \rightarrow 5 \text{ because } 5(3) \rightarrow 15$$

In due course, when the teacher is satisfied that the relationship between division and multiplication is clearly understood, she can allow the child to record the division part of the statement without adding the explanatory multiplication clause.

It is important for the children to become familiar with the idea of dividing by *grouping* as well as by sharing: that is to say, if there are six sweets how many children can have two each? Appropriate assignments and practical activities should therefore be devised by the teacher to give the children experience of this form of division. Those using Cuisenaire rods will have encountered it quite naturally: how many red rods are the same length as a dark green one? Since these children are by now reading fairly well, the number of words needed to describe such an assignment presents no great problem.

### FRACTIONS

The most familiar fraction to young children is the half. Indeed many of them use the word and understand that it means some kind of 'fair share' long before one introduces activities which have as their objective the formalisation of the child's understanding of halves and quarters. Sooner or later the formalisation becomes necessary, because as mathematics progress fractions are used. In learning to tell the time, for instance, *half past* must have more than merely linguistic meaning; so must *halfpenny*, which will eventually be used in the classroom shop.

So there comes a time when fractions should be systematically introduced, and for some children this is by about the middle infants' stage, or a bit earlier. For practical purposes those which are most needed are $\frac{1}{2}$, $\frac{1}{4}$ and $\frac{3}{4}$, and they should be presented in that order.

They offer a useful preparation for division because they include the principle of *equal sharing*. Discussion is needed to establish the fact that though an apple, for example, may be divided into two parts (or *fractions*), it must be equally divided if the two parts are to be *halves*; otherwise Linda will have a larger fraction of the apple than Ian.

One way of introducing this principle clearly is for the teacher to work with a group of children, who are given gummed paper squares, scissors, and plain paper or a special 'Book of Fractions' in which they will, over a period of time, assemble fractions of various kinds. The children may begin by cutting off two strips of paper roughly as wide as a ruler, and one of these strips represents the *whole*. How should the second strip be folded in half? (Let the children determine how to do this.) The strip is then cut at the fold, and the children stick the two halves in their books, beneath the one whole. They have established that each shorter strip is a *half* of the longer strip, and that there are two halves in one whole.

This should, on subsequent occasions, be repeated with a variety of shapes – circles, squares, rectangles – with numbers of objects, with coins, by splitting sets, and by any other practical activity that lends itself to the purpose.

The children are taught to recognise, and to write, the words *half* and *halves*, and they also need to be given the shorthand written form. But this must be explained to them. The half is one of two equal parts into which the whole has been divided; it is therefore written as $\frac{1}{2}$.

Experience, in which discussion plays a particularly vital part, is necessary to enable children to abstract the principle that if two threes are six, three is a *half* of six; this helps to prepare them for a later stage of progression, when they will find that if any whole is doubled, the original whole then becomes half of the new whole.

Next, $\frac{1}{4}$ and $\frac{3}{4}$ should be presented in the same practical way as $\frac{1}{2}$. It is important to remember that though these fractions may be introduced in a systematic way in order that the children shall understand what they mean, their *application* should not at this stage be regarded as an isolated mathematical activity. Children should meet them in the course of their everyday practical experience, so that the words become part of their vocabulary (in a mathematical sense), and so that they can absorb the principle securely enough to give meaning to such terms as $\frac{1}{4}$ past 3 and $2\frac{1}{2}$p. Anything more than this properly belongs to a later stage than that with which we are now dealing.

## THE COUNTING SYSTEM

During this stage the counting system should be extended towards 100. This does not mean that we expect the children to understand conservation of such large numbers. But they should become aware that there *is* a system, that it has a form and pattern, that the numbers signify progressively larger quantitative values and that they have graphical forms which the children learn to recognise.

With the advent of metrication, children meet some of the larger numbers, even those beyond 100, much earlier than they used to.[7] It is probable that in activities like weighing these large numbers may well be used as names long before they can convey to the children an accurate notion of numerical quantity. However, they will eventually have to be used in a quantitative sense, and in the meantime children can be prepared for the later use of the larger numbers by being equipped with an understanding of the counting system.

So they should learn to count to 100, to count in tens and twos, to go as far as they can in recognising and writing them, and to understand the difference between odd and even numbers. A hundred chart is now an invaluable aid in helping children to see the pattern of the number system. This can be duplicated on expendable sheets, and the children asked to make patterns of different numbers by colouring the appropriate squares, e.g.

A pattern of 3 s

| 1 | 2 | 3 | 4 | 5 | 6 | 7 | 8 | 9 | 10 |
|---|---|---|---|---|---|---|---|---|---|
| 11 | 12 | 13 | 14 | 15 | 16 | 17 | 18 | 19 | 20 |
| 21 | 22 | 23 | 24 | 25 | 26 | 27 | 28 | 29 | 30 |
| 31 | 32 | 33 | 34 | 35 | 36 | 37 | 38 | 39 | 40 |
| 41 | 42 | 43 | 44 | 45 | 46 | 47 | 48 | 49 | 50 |
| 51 | 52 | 53 | 54 | 55 | 56 | 57 | 58 | 59 | 60 |
| 61 | 62 | 63 | 64 | 65 | 66 | 67 | 68 | 69 | 70 |
| 71 | 72 | 73 | 74 | 75 | 76 | 77 | 78 | 79 | 80 |
| 81 | 82 | 83 | 84 | 85 | 86 | 87 | 88 | 89 | 90 |
| 91 | 92 | 93 | 94 | 95 | 96 | 97 | 98 | 99 | 100 |

Other uses of a hundred chart are:

1. Counting in twos, tens, fives, etc., colouring the squares that are counted.
2. Colouring odd or even numbers.
3. Colouring every tenth square, starting at four (this helps towards an understanding of tens and units).
4. Colouring alternate horizontal rows green and yellow. How many rows? How many squares in each row?

There may be one or two advanced children who, after a while, can assemble their own charts by writing the numerals on plain squared paper, and then making up their own patterns.

Pegboards and coloured pegs can be used in the same way. For most children, the numerals would need to be painted above each hole. A small piece of wood glued under each corner of the pegboard raises it enough for the pegs to hold.

Oral counting in odd minutes is useful, and games of 'finding thirty-two' (on the hundred chart) or 'finding the number that is one more than forty-six'. The purpose of all this is not, we repeat, for children to *operate* on these larger numbers; merely for them to become familiar with the number system and with the way in which it is constructed. We would suggest that at this stage the children should not be introduced to any number base other than ten. The foundations will be sounder if they are not spread too widely.

# NOTES

1 Operations are *commutative* if the order of the elements does not affect the result. Subtraction is not commutative (i.e. $4-3$ is not the same as $3-4$); neither is division, but multiplication is. Many people use, probably unconsciously, the commutative laws of addition and multiplication for quick calculation, so its introduction at an early stage of mathematical learning is material to the later application of mathematics in daily life.
2 For ideas on some more unusual addition games, see especially Harold Fletcher (ed.), Teacher's Resource Book, *Mathematics for Schools,* Level I (London: Addison-Wesley Publishers Ltd, 1970).
3 E. M. Williams and Hilary Shuard, *Primary Mathematics Today* (London: Longman, 1970), p. 79.
4 *Mathematics in Primary Schools,* 4th edn. (London: HMSO, 1972), p. 22.
5 Ibid.
6 Ibid.
7 250 grams, for example, is just under 9 oz; but 250 is a very large number compared with 8 or 9.

*Chapter 13*

# Other Mathematical Activity

## LENGTH AND AREA

When children can conserve number, activities involving iteration can begin; that is to say, the child can learn to take a measurement by repeating a unit and counting the number of repetitions.

As in the earlier stage, the children should start by using parts of their bodies. A chalk line can be drawn on the floor, and the teacher and some of the children see how many of their feet it takes to go from end to end of the line. Similar measurements might be taken with paces, handspans and so on. Provided there is enough opportunity to talk about these activities, so that views can be exchanged among the children and with the teacher, the advantage of having a fixed unit of measurement will sooner or later become apparent.

The question then arises of what kind of fixed unit is the best to select. In the days before metrication, the foot ruler was good because it was a handy size. It was neither too long for small children to handle easily, nor so small as to make it difficult for them to iterate without so great a margin of inaccuracy that the benefit of having a fixed unit was obscured.

The fixed unit with which the children begin need not, of course, be related to any standard unit of measurement. Sticks of equal length will do, or reading books, or milk straws; and many teachers like to use these as the first stage in the process of measuring with a particular unit. There is something to be said, however, for moving on fairly soon to something which, though the children do not yet know it, is in fact a standard unit of measurement. They will eventually, at a later stage, measure in metres and centimetres; and they will have something of a start if they are already accustomed to a unit which resolves itself easily into the measures they will then be using.

Metres and centimetres are, however, very large or very small. Since we would all agree that centimetres are out because at this stage they are much too small, we seem to be left with the metre stick – and some teachers prefer to start children off with these. However, it is difficult when you are not very large yourself to keep a thing of that length under control as you move it from one place to the next, and you feel a bit sore if you are blamed because you hit an innocent bystander in passing; and by the time you have managed to re-position it, you have forgotten the last number you counted anyway. A further difficulty is finding enough surfaces to measure when the unit is so long.

A workable compromise is to use the orange Cuisenaire rod, which is ten centimetres long. It is a little too small to be ideal, but not so small as to disturb unduly the measuring activity. At this stage, it is just called the 'orange rod', with no measurement name attached. The line on the floor is found to be so many 'orange rods long', or the locker is 'six orange rods high'. If the classroom is not equipped with Cuisenaire apparatus, ten centimetre units can very easily be made from the beading that is sold in do-it-yourself shops, and if the teacher wants to make them more colourful a large felt-tipped pen will do the job in moments.

As the children's skill at handling the unit of measurement increases, they will iterate more accurately. It helps if they work in pairs – one to mark and one to measure. The question of vocabulary may begin to present some problems. Words like *broad, wide* and *deep*, which are not precisely interchangeable, can be most confusing unless the teacher is careful in the use of vocabulary and in explaining what the words mean.

Children now need plenty of experience in practical measurement, but it is not very satisfactory just to let them loose to 'do some measuring'. Some of them will have no idea of what they might measure; others will try to measure something quite unsuitable like the height of the door; and those who do manage to measure the shelf will not know what to do with the measurement once they have taken it.

It is not always practicable – nor is it desirable – for a teacher to break away from whatever she is doing to instruct a pair of children, or a group, about what to measure. So she may delegate this duty to practical work cards, which give the children ideas about this and about how to record the measurement. The cards should allow for the fact that the reading and writing skills of most of these children are still pretty limited, and they might be designed in this way:

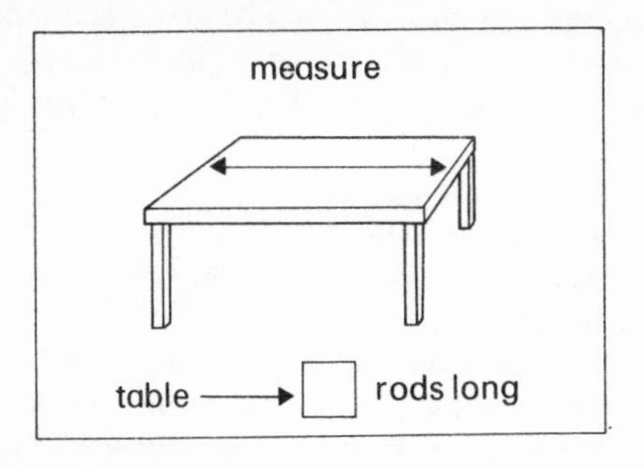


The children should be shown how to measure round their waists or heads with a piece of string, and then to take the measurement by measuring the string. Brightly coloured cardboard strips of varying lengths can be pinned up at child height in different parts of the room, with instructions to measure them included on the measuring cards. Tissue paper strips of different lengths and colours can be stapled on to a large piece of card, i.e.

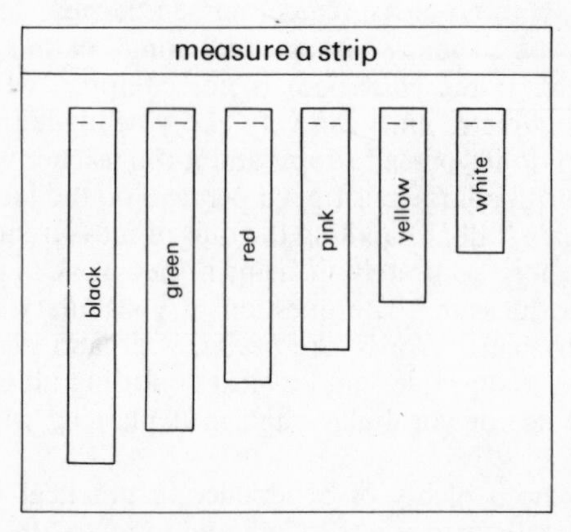


This piece of apparatus is also useful at a later stage, when accompanying cards can ask such questions as 'How much longer is the pink strip than the yellow one?'

Many objects in the room will not measure exactly in the unit of the rod, and after a while children will begin to question what to do with 'the bit left over'. In the past, this was the moment for introducing inches. But whereas '1 foot 4 inches' was a sensible measurement, '3 rods and 4 centimetres' is not.

It is sometimes suggested that children should now use some of the other rods, selecting the one that will fill the remaining space. This works quite well if the children are measuring a surface that lies flat, like a book or a table top; but it takes a good deal of dexterity to do this with, for instance, a blackboard or the leg of a chair. We suggest that it is better, at first, for children to use the formula '3 rods and a bit' or 'nearly 6 rods'.

One should not exclude the use of the larger unit – the metre stick – in suitable conditions; that is, where there is enough space, and where the teacher is satisfied that the activity can be sufficiently controlled to be mathematically beneficial.

When the children progress to an understanding of the larger numbers, and when their muscular co-ordination is developed enough to enable them to handle a ruler with the accuracy required for a very small unit, they can begin measuring in centimetres. The measurements they are asked to take should be no more than about 20 cm at first. Work cards would therefore suggest measuring such things as a book, a shoe, a hand, a wrist; the sides of a picture pasted on a card; strips of gummed paper of different lengths and colours, pasted on a card.

Measuring activities can give the children valuable experience in discovering something about the properties of shapes. For example, a square can be drawn on card, each side and each diagonal being shown in a different colour. An instruction to measure the blue line, the green line, the yellow line, etc. will help the child to discover that the sides of a square are of equal length, as are the diagonals; similar experience with triangles of different size and shape will show them that triangles may have three sides of equal length, or only two, or none at all. This kind of activity, of course, is of far more mathematical value if immediate discussion can take place about what the child has discovered from his measurements. A useful way of ensuring that this happens is for the teacher to include on these cards a symbol of some kind – such as a star in the corner – which is known to mean that the child must see the teacher as soon as he has completed the card.

The children should, in time, begin *constructing* measurements of length. They can draw lines so many cm long, and also pictures; a fish 12 cm long, a tower 14 cm high. A valuable activity is for the child to have squared paper (1 cm square) on which he colours lines of equal length, beginning each time on a different square (which the teacher has marked with a dot).

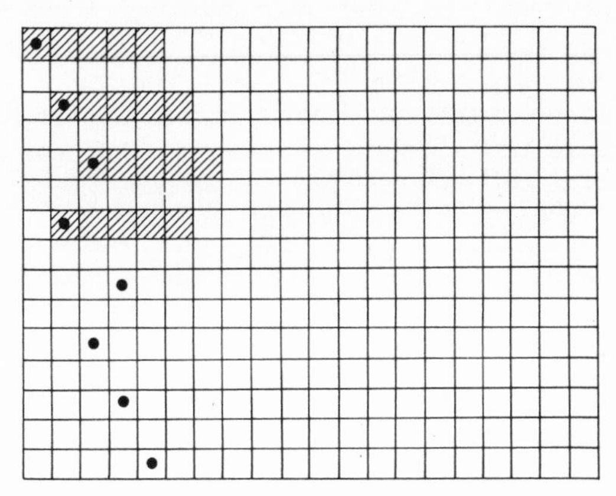

This helps the child towards an understanding of the conservation of length, because he begins to see that though some lines protrude beyond others, this does not make them longer. The squared paper helps him in this context, to confirm the equivalence of the lengths he is constructing.

If a child measures the length of his table with rods, a natural extension of the activity is for him to see how many cards or books (of identical size and shape) will cover the table's surface. If the floor is tiled, he can put a chalk mark on each tile that is enclosed by a space 3 tiles along and 3 tiles across, and see how many tiles have filled the space. On squared paper, he can make patterns, using a given number of squares and arranging them in any way he likes.

These are not activities directed at 'finding the area of . . .', for it would be much too soon for this. The object is for the child to have experience of assembling and arranging two-dimensional shapes, in preparation for the time when he will approach area as a mathematical activity of a different kind. If two or three children can engage together in these early experiences of area, comparing the patterns they have made and discussing how they might invent some more, they are likely to extend both their experience and their language to a greater degree than if they were always to work individually.

## SUBSTANCE OR MASS, AND LIQUID MEASURE

Piaget's view that the concept of the conservation of substance seems to develop much later than was once believed has been substantially confirmed by subsequent research. However, we have noted that there is one aspect of the conservation of substance in which more recent investigations have produced results that are somewhat at variance with Piaget's findings, and this point has particular application to the stage with which we are now concerned.[1]

Piaget appears to suggest that once a child can conserve substance he is a conserver with all materials, whether continuous or discontinuous; and that with most children this understanding does not develop until after the age of about eight years. But Lovell reports a number of investigations which indicate that, as far as continuous quantities are concerned, children who are conservers with one material are not necessarily conservers with another.[2] They may, for example, have grasped the concept of conservation with water, but not with plasticine. The argument that these children might be in the Piagetian transitional stage, when conservation is still unreliable, does not apply. Lovell and others found that children who were conservers with water, but not with plasticine, had a reliable grasp of conservation in regard to water. It did not come and go, as one expects in Piaget's transitional stage. And yet the same children really had not grasped the concept of conservation with plasticine.

The evidence suggests that an understanding of the conservation of liquid may arrive somewhat earlier than that of other continuous substances, and that as far as material like plasticine or clay is concerned, only a minority of children are reliable conservers even in the first year of the junior school. This being so, we should not expect more of the younger children than activities which continue to be experiential and exploratory. We should defer the measurement, even of liquids, until 'top infants', and, for most children, the measurement of other kinds of substance until even later. We should therefore retain substantially the same approach as was suggested for the earlier stage, adding variety of experience where we can, but not asking too soon for definitive measurement or written recording.

## WEIGHING

One way of giving the child additional experience which will lead him towards an understanding of the conservation of substance is to give him experience of 'weighing' – or, more accurately, of *balancing* – a variety of materials. In the earlier stage he was introduced to the balance, with activities that enabled him to observe its behaviour and its relationship with the notion of *heavy* and *light*. He saw what happened as a result of his actions when he put more conkers in one pan or more shells in the other. With this experience behind him he is ready to go a step further, and engage in balancing activities in which he balances objects against standard units.

We should be clear about the fact that it is not yet our intention to ask for a definitive measurement of substance, as expressed in terms of weight. Our objective is to give the child the opportunity of recognising that only a certain amount of material will balance against a standard weight; that it takes more of some kinds of material to do this, than others; and that if it takes more it always takes more. This is the kind of foundation experience that is necessary if the child is later to arrive at the stage when he can really understand the more complex principles of balancing. He should learn, for example, that if each of two different kinds of material balance against the same standard unit, they will also balance against each other; that if a ball of plasticine balances a standard unit, and still does so when it has been rolled into a sausage, the total amount of plasticine appears not to have altered; and that standard units can be given names, which are convenient for describing different quantities.

The more advanced children may reach this stage before the end of the infant school, but there is no point in rushing one's fences and putting mathematical understanding at risk. We should therefore give all children a good deal of preliminary experience of balancing objects against standard units, before asking them to treat such units as sophisticated indicators of measurement.

At this stage we need a unit that does not suffer from the disadvantages of grams and kilos. The gram is so small that it takes a very large number to give the kind of weight that is practicable for a young child to manage; and these children are not yet at home with such large numbers. A kilo, on the other hand, is a comparatively heavy weight, and this limits the range of articles that we can give the child to use in the classroom with the balance. One shrinks from the thought of the number of acorns required to balance two kilos!

One solution is to use 'home-made' units which do not have these drawbacks. It is useful to have two contrasting units – called *big weights* and *little weights*. These can be made from small bags filled with gravel. It helps if the bags are of two different colours, because this makes identification on work cards easier for children who cannot yet read very much. We suggest that the big weights might weigh about 200 grams and the little ones about 50 grams. This allows for light objects like beads to be balanced against the little weight, without there having to be too many of them; and yet it is possible to provide a reasonable range of objects to balance the heavier weight – marbles, stones, wooden building bricks, Dinky toys, cotton reels, etc. There is also enough difference between the two weights for them to be distinctive. Three or four weights of each kind are needed for use with one pair of balances.

As always, there is nothing to beat direct contact between the teacher and a small group of children for the activity to be of maximum benefit, and the teacher should try to set aside whatever allowance of time is possible for introducing it and for keeping in touch with it as time goes on. Arrangements are, however, necessary for giving the children additional practical experience, and the assignment card is useful for this purpose. A few examples of those which can help the child to make progress with his weighing activities are given below.

1.

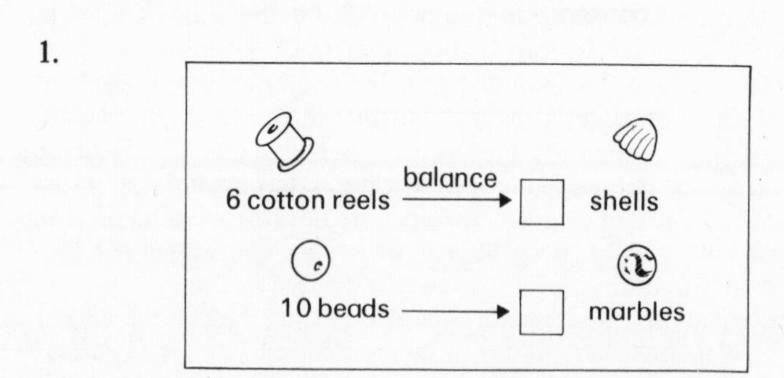


When recording, the child copies the writing and inserts the number in the square.

2. Make up some parcels in each of 3 weights: in red paper weighing 200 grams, in blue weighing 100 grams and in green weighing 50 grams.

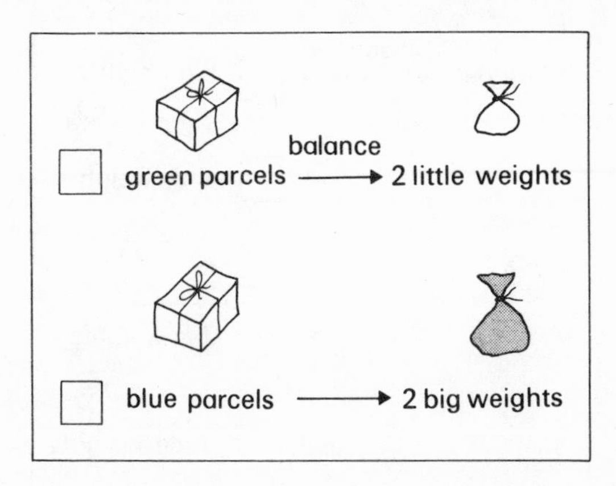


(The big weights might be in brown bags and the little ones in yellow. The colours on the card will help the children to identify them.)

3.

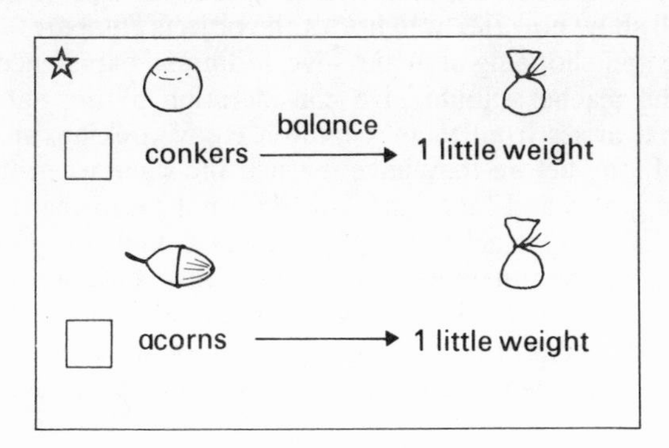


This card has a star in the corner to indicate that the child must see the teacher as soon as he has completed it (see suggestion on p. 137). The reason in this case is so that the teacher can now ask the child to balance the conkers against the acorns and see what happens. To explain this on the card would take too many words, and the child would not be able to read it.

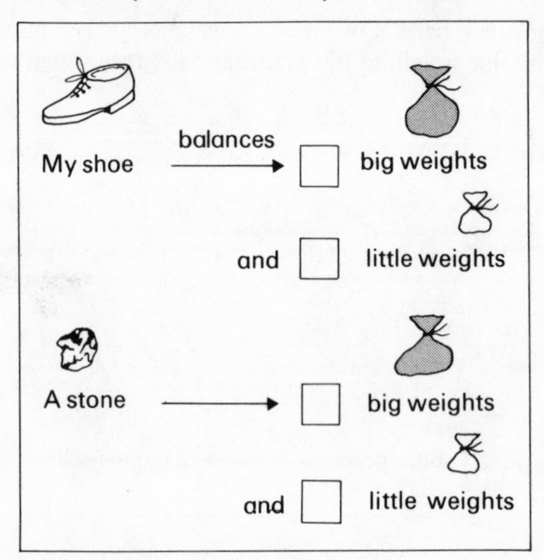


As reading advances, less illustration is required.

Another way of recording balancing, which is best undertaken as a small group activity with the teacher, is by means of a block graph. Units such as shells are chosen, and various objects are balanced against shells. The shells required for each object are entered in the vertical columns, the children drawing the requisite number on sticky paper and assembling them after each balancing operation. The follow-up discussion will show how the 'weights' of the objects compare.

Cooking and shopping activities give additional experience of weighing, but the teacher should give consideration to the nature of the learning that arises from them. Children enjoy cooking, and often do it in school long before they have reached the stage when they understand what grams and kilos are. This does not mean that they should not cook. But the teacher should recognise that weighing out ingredients and giving them conventional weight names is, at this stage, an *experiential* exercise. It is valuable as a mathematical preparation, but it does not necessarily indicate an understanding of the quantitative value of the names which are applied to measurements made by means of weighing.

With regard to shopping, it is suggested that in view of the large numbers required for metric weights, their introduction should be delayed for longer than used to be the case with ounces and pounds. If there should be any children who are ready to deal with these units later in the infant school, they should begin with only a limited number of the normal range of units. The normal range is 5g, 10g, 20g, 50g, 100g, 200g, 500g and 1kg; of these, we would suggest introducing no more than the 50g (just under 2oz), the 100g (nearly 4oz), the 200g (just over 7oz) and – just possibly – the 1kg. Very clear explanation of the meaning of the weight names is necessary. Provided, however, that the child has had plenty of experience of balancing materials against 'big weights' and 'little weights', the activity of weighing 100g of sweets can begin to acquire mathematical meaning. But there is nothing to be gained, and possibly a good deal to be lost, by introducing metric units too soon.

Although the kilogram is, in fact, a unit of mass and not of weight, we do not attempt to make the distinction in the infant school (or, indeed, for a long time afterwards). In any case, kilograms are used to describe units of weight in everyday life, so we are not doing anything very reprehensible in applying these words to weighing activities when children are ready to assimilate them.

## TIME

Towards middle infants, some children may be ready to begin taking further steps in regard to the question of time. These steps concern two aspects of learning:

1. Moving towards an understanding of the passage or duration of time.
2. Learning to read the clock face.

### *The Passage or Duration of Time*

Our aim here is to give children the kind of experience which will help them to make the connection between the passage or duration of time and the way in which it is conventionally measured. In view of the extremely abstract nature of the subject we have to find some concrete means of achieving this, and as a first step one of the best ways of doing it is by using a water clock.

A simple version of this is to have a glass jar with a tin on top of it: a hole is pierced in the tin, so that water can drip slowly into the jar.

After the necessary introductory discussion, the clock is set up at the beginning of an activity that will last no more than about ten minutes or so (such as the class milk time). When the last bottle of milk is put away a mark is made on the jar with a felt pen, at the water level.

A second activity, which is likely to take noticeably longer, is then measured in the same way. This might, for example, be playtime. A discussion then takes place about the fact that *more* water dripped into the jar during playtime than milktime, which should lead to a consideration by the children of which *took longer.* Further activities, at fairly frequent intervals, are measured in the same way. (If too much time elapses between these measurements, the children will forget about them and the significance of the comparisons will be lost.)

At this stage there will be no formal recording of these operations by individual children, but an informal class record may sometimes be useful. A sheet of paper may be pinned up, on which is written 'milk time takes longer than playtime'. An arrow might indicate 'takes longer than', so that further additions to the record require less writing. A second piece of paper would record activities that take less time than others. If the children's writing skill is still very limited, the teacher should write the words for them – with their participation in assembling the statement.

It may be helpful to mention one or two practical points in connection with making a water clock. It tends to be a bit top heavy when there is water in the tin, so the jar should have a firm base and the clock should be situated in a place where it will not be easily knocked over. (A fixed work top, for instance, is safer than a free-standing table which may be inadvertently bumped.) The tin, ideally, should fit neatly into the top of the jar. The teacher should experiment with the clock before introducing it to the children, to make sure the hole in the tin is neither too large nor too small for it to work effectively. The jar should always have a greater capacity than the tin, in case some enthusiast decides to set up an experiment on his own and then forgets about it because his attention is caught by something else.

Apart from water clocks, there are other timing devices that are sometimes used in infant schools. These include egg-timers, minute timers, stop-watches and candle clocks. These devices, however, are more appropriate to later stages. An egg-timer could possibly be used in the same way as a water clock, but only in connection with very short activities; and there is hardly enough distinction in timing to help children at this stage to make comparisons. Later on, egg-timers can be used for timing games and so can minute timers and stop-watches. But if these activities are to promote learning about time, the child's understanding needs to be more accurate and sophisticated than we can expect it to be at the beginning of time activities. Candle clocks are more useful at a more advanced level as well, because their operation is not at first as clear to children as is that of water clocks. A lighted candle measures the passage of time by disappearing, whereas with a water clock the water is there for the child to see. The teacher should, however, keep these other devices in mind for use with the more advanced children in top infants.

*Reading the Clock Face*

This, of course, is simply a matter of reading a dial, and it is a useful accomplishment which need not be delayed until the child really understands abstract principles about the continuity of time and the nature of its measurement. A householder, after all, may read an electricity meter correctly without having any understanding at all of watts and volts.

This is not to say, however, that it is quite easy for all children to learn to read the clock face, or that certain maturational and mathematical prerequisites are not necessary before they can begin to do so. The minimum mathematical requirement is that the child must have a clear understanding of numbers to at least twelve, and he must know what halves and quarters are. He must also be maturationally capable of recognising that when the dial shows 12 o'clock, it is registering the 'twelve o'clock' that he has often heard used to describe dinner-time. The connection is not as obvious to the child as it is to the adult, and until he has made it there is no point in asking him to learn about the dial.

For teaching children to tell the time, the equipment needed is:

A large, clear clock face with movable hands.
A few small clock faces which the children will use.
A rubber clock face stamp.
Some cut-out manilla clock faces which have been stamped.
A real clock, *not* with Roman numerals. (The classroom is an excellent resting place for the old kitchen clock that is no longer worth repairing.)

Ideally, there should also be on the wall of every classroom a real clock that goes. Regrettably this is not often provided in infant classes, but the teacher who has one is fortunate.

*Introducing the Clock Face.* Once again, this can most advantageously be done with a group. By now the children concerned know what 'telling the time' means. The introductory discussion should centre on the appearance of the clock face, the numerals that are shown on it, the way in which they are positioned and the fact that there are two hands, one being longer than the other. The large clock face is used for this purpose. The next step is to turn the hands of the old kitchen clock, while the children watch to see what happens; and in the course of discussion it is established that while the big hand moves all the way round the clock face, the little or 'lazy' hand moves only from one numeral to the next.

When this basic principle of the movement of the hands is recognised, the children can move on to learning to tell the time at the hour. At this point it is worth explaining to them that o'clock is simply an abbreviation of the old term *of the clock* and the apostrophe is a substitution for the letters which have been omitted. (Despite the explanation some children will at first use a comma in recording instead of an apostrophe, and the teacher must help them to get it right.)

It must then be explained that the big hand on the twelve always indicates that it is *something o'clock*, and the little hand shows what the *something* is. The children in the group take turns at arranging the hands on the large clock face at various hours. It seems in practice to be easier for children if they always position the big hand before the small one.

On a subsequent occasion, the children may each be given a small clock face on which they position the hands at different hours and the others in the group tell the time that each child has chosen. This helps to establish familiarity with the clock face and with positioning the hands at the hour.

The children should now be shown how to do this on paper. They should each be given a sheet on which some clock faces have been stamped, and they draw the hands to indicate 2 o'clock, 7 o'clock, etc., writing the time name beneath each clock face. It is important to ensure that from the beginning children show clearly the difference in the length of the two hands; otherwise it can become very confusing for them, particularly when they move on to showing other times besides the hour.

Sometimes they should be asked to draw their own clock faces and write in the numerals, instead of having them already stamped. This helps to fix the clock face in the child's mind. However, it is unsatisfactory to do no more than ask him to draw a circle and put in the numerals, because his clock face will almost certainly look like this:

1 o'clock

This is where the stamped manilla clock faces are helpful. The child uses one as a template to draw the circle, and is shown how to position the template and make a mark at 12, 3, 6, 9; these numerals are then written in, and the others positioned between them. Alternatively the child can first make a mark at every numeral. The important thing is that he learns from the beginning to put the numerals in the right places.

Assignment cards should give the child practice at recording the time that is shown on a clock face, and also the reverse process of positioning the hands at certain given times. Care should be taken to ensure that the wording on such cards in minimal, and repetitive, because some of these children still have rather limited reading skills.

If there is a real working clock in the room, the teacher should take the opportunity now and again of drawing attention to it at round about the hour, asking a child in the group concerned what time the clock says. It will then be very natural to introduce such phrases as *nearly 10 o'clock* and *just after 2 o'clock*, which paves the way for the time, at a much later stage, when the advantage of sub-dividing the hour into minutes becomes apparent. If there is no working clock in the room, the teacher can sometimes do the same thing with the old kitchen clock or even the artificial clock face.

After the hour, half past, and then quarter past, should be introduced and followed up in the same way. It is important that each time the children should observe, and discuss, what happens to the small hand at the half and quarter hour when the movement is shown on the old kitchen clock. They will also need help in positioning the small hand correctly when drawing it on a clock face.

Sometimes $\frac{1}{4}$ *to* the hour presents a little more difficulty because, although it is logical to say $\frac{1}{4}$ *past* and $\frac{1}{2}$ *past*, it is a bit less logical to say $\frac{1}{4}$ *to*. The teacher's best recourse is to describe this time-name in linguistic terms, explaining that as it is a quarter before the next hour, it is more convenient to call it $\frac{1}{4}$ *to* than $\frac{3}{4}$ *past*. Bearing in mind the understanding of halves and quarters that the children should have by now, the language problem should not prove to be too great.

Wall charts are helpful when children are learning to read the clock face. These show

1. A plain circle to indicate the hour.
2. A circle, of which half is coloured when the half hour is introduced.
3. Another circle, with the first quarter coloured for $\frac{1}{4}$ past.
4. A fourth circle, with only the last quarter segment left uncovered, for $\frac{1}{4}$ to.

At this stage, the symbols $\frac{1}{2}$ and $\frac{1}{4}$ should be used in recording, rather than the words *half* and *quarter*. This will ensure that the mathematical connection is made with the activities on fractions which will already have been undertaken, and which will be continuing. The forms *4.15, 4.30* and *4.45* should not be introduced until much later, when the children have reached the stage of being able to divide the hour into minutes. This, however, together with matters such as the use of Roman numerals, the 24-hour clock, timing games and so on, do not enter into the learning of children at this present stage. The water clock can, however, be used to measure the passage of quarter and half hours when the children are able to read the clock face at these times.

## MONEY

By now, we may make certain assumptions about the level of understanding the children concerned have reached. They can recognise the

1p, 2p, 5p and 10p coins. They can understand the conservation of number to at least ten. They understand that shopping is a transaction and they know how to use the shop. They are ready to leave the playshop behind and move on to something which will extend their learning.

It is time to introduce a different kind of shop, partly because some of the objectives of the shopping activity have changed and partly because the old shop has had a long enough life anyway. The objectives of the shop now are:

1. To help children to extend their understanding of money value.
2. To give them additional experience of the processes of addition, subtraction and multiplication.
3. To give them further opportunity for making social and intellectual contact with each other, in a way that is both enjoyable and educationally beneficial.

If the first two objectives are to be reached, certain of the shopping activities must now be structured with these ends in view. The third is part of the broader classroom scene, but it applies to shopping activity as it does to any other.

The question of realistic pricing is still something of a problem, for the arguments have not changed much since the first play-shop was established. One way of preventing it from being too troublesome is to stock the new shop with things which the children make, because these have no counterparts with goods that are found in the shops outside. A toyshop is one possibility, so we will use it as an example though of course there are other alternatives. We will also bear in mind that no classroom shop has an unlimited life, and if interest in shopping is to be sustained throughout the infant school years a different kind of shop will need to replace the existing one every now and again.

*The Toyshop*

After a preliminary discussion about the enterprise, the children, in the course of perhaps a couple of weeks, make an agreed range of merchandise. The toys must, of course, be simple to make, reasonably durable and easily replaced. There should be no more than about six different toys to begin with, and there should be several of each. To take a few random examples, they might be windmills, flags, simple boats with wooden or even match-box hulls, papier mâché balls, strings of beads made from flour and salt dough and some puppets (which can be 'bought' from the shop when they are needed for other purposes).

The children have a firm stake in the shop since they are manufacturing the goods for it, and this will undoubtedly give them an added interest in it when it is finally assembled. They should be helped to display the goods attractively and accessibly, and there should be a 'till' (which need be no more than the lid of a box divided into compartments), shopping bags, purses or tins for the shoppers and a shopkeepers' list as suggested on page 80.

It would be impracticable to mark each item with its price, so a price list should be displayed by the shop. Because of limited reading skills, the price list should show the name of each item, a picture of it, and its price.

Those children in the class who are still at the earlier stage should be allowed to use the new shop in the same informal way as they did the play-shop. However, one sometimes finds that some of them also want to do 'proper shopping' like their friends and they should be permitted to do so; but the teacher would not place any emphasis on the accuracy of the transaction or let them feel that they must not help themselves to a shopping card if they want to, because they cannot yet add 2p and 3p. If they enjoy the experience they will learn from it, and from each other. Moreover, the children who will now do more structured shopping should also be allowed to shop freely at other times if they wish, as long as there is room for them.

Provided they supplement and do not replace contact and conversation with the teacher, shopping cards are now useful for the children who are ready to move ahead. The progression that can be expected is as follows:

1. The addition of two separate items to a total of up to 10p, without having to get change; payment would therefore be made in 1p, 2p and possibly 5p coins.
2. The addition of two or three separate items to a total of up to 10p, with change (subtraction); payment in 5p and 10p coins.
3. The inclusion of two or more identical items (multiplication); the total may go to 12p or 15p and payment made in combinations of the four coins used (1p, 2p, 5p, 10p). The children now decide upon the combination of coins they need, and get change if necessary.
4. The extension of the total to 20p, with several items to be purchased, both separate and identical.

Until reading skill is sufficiently advanced, the items to be purchased must be illustrated on the cards and words kept to a minimum, e.g.

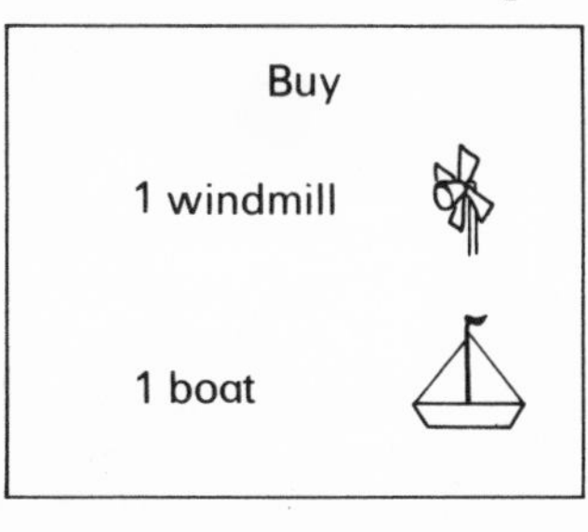

At stage 2, when change is introduced, the coin or coins to be taken to the shop should be shown on the card, e.g.

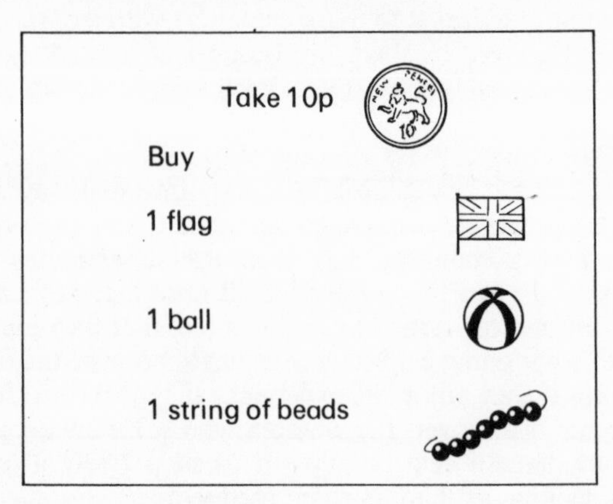


At the appropriate stage, measuring can be included in the shopping activity. Small rolls of coloured crepe paper, of the kind which is used to make streamers, make excellent 'ribbon'. The unit of measurement may be a rod or a metre stick.

The inclusion of weighing as a structured activity in the shop is probably better deferred until a much later stage, when the children can weigh in conventional units. However, there is no reason why the 'big weight' and 'little weight' system should not be used in the meantime, in informal shopping activities, if the teacher wishes. It is just rather clumsy to include on shopping cards.

The $\frac{1}{2}$p has not been suggested as one of the coins for shopping at this stage, even though the children are engaging by now in other activities concerned with fractions. The reason for this is that in these other activities fractions are being explored with concrete materials. Even in the case of time, half an hour can be visually associated with half the clock face. The $\frac{1}{2}$p, however, is an indication of *value*, which is somewhat more abstract; and in shopping it would involve the addition and subtraction of halves, which we could not yet ask children to do. We therefore exclude it at this level of dealing with money, though many children will certainly have heard of it and it may be conversationally used as a money name.

*Recording Shopping*

There is something to be said for giving children a special book in which to record their shopping transactions. This can be made from kitchen paper, and need be no larger than is necessary to record one transaction on a page. On the front of the book is written 'I spent', so that the words are at hand for the child to copy. Having bought his windmill and his boat, he records

| | |
|---|---|
| windmill | 2p |
| boat | 5p |
| I spent | 7p |

The words for his purchases are copied from the shopping card. When he reaches the stage of getting change, the teacher ensures that he knows what the word 'change' means and adds it to the cover of his shopping book. Without the help of these words, and a simple formula (which the teacher has shown him) for recording the transaction, the recording of shopping activities can become very confused and laborious for the child and its value is largely lost. As he makes progress in recording generally, he can record shopping computations in ways that are similar to those he uses in his other number work.

*Desk Shops*

If space is so limited that it is impossible to have a classroom shop, desk shops are an alternative. They are very much second best, but they are better than nothing.

A desk shop is a shoe box containing small items for sale, and is used by two to four children at their tables. One might, for example, be a baker's shop. Miniature loaves, rolls, buns, cakes, etc. are made from flour and salt dough. A coat of clear varnish applied after painting makes them much stronger. A price list is kept in the box, and if it is made from a piece of manilla folded in half so that it will stand on the table, the prices can be shown on both sides and the shop used simultaneously by children who are sitting opposite each other.

Shopping cards can be made for the desk shops and the children pay for their purchases by putting their coins into a common till and collecting their change.

Desk shops do not offer the variety of experience that is possible with a classroom shop, but they give children more interesting opportunities for making money computations than they can get from straightforward computation cards. For this reason they are worth considering if a better shop is out of the question.

*Wall Shops*

These do neither more nor less than provide further variation in opportunities for making money computations. A wall shop is a large piece of manilla, divided into sections, with the goods for sale illustrated and priced.

| The fruit shop | |
|---|---|
| apples<br>3p each | lemons<br>6p each |
| pears<br>4p each | bananas<br>5p each |
| oranges<br>7p each | plums<br>2p each |

The 'shop' is pinned to the wall, and accompanying cards give experience in computing money.

*Coin Cards*

Another way of giving practice in computational skills is by using cards to which groups of paper or cardboard coins are glued. On each card there may be four groups of the denominations in use, assembled in different ways. The child calculates the total value of each group. Subtraction may be added by stating a sum from which the value of each group is to be deducted. Other cards may show sums of money. which the child assembles with various coins and records the coins he has used.

The principle of sharing a sum of money may also be included. For this, a card would show pictures of perhaps four children, and the child is asked to share 12p equally between John, Mary and Janis, Mary and Keith, and so on, stating how much each will have, and why.

Activities of this kind give variety in practising computational skills, and as long as the teacher recognises that they are *practice* activities, and does not allow them to replace those that are experiential, they have a useful part to play in helping children to learn about money.

NOTES

1 See page 68.
2 K. Lovell, *The Growth of Basic Mathematical and Scientific Concepts in Children* (London: ULP, third impression, 1971), pp. 64–8.

PART FOUR

# Organisation and Management

*Chapter 14*

# The Children, the Teacher and the Materials

We have long since recognised that for children to learn pleasurably and to their advantage, they have to be freed from the constraints imposed by the class teaching of the past. We know that not every child can benefit from the same teaching at the same time, and that if children are to learn with understanding they must be able to participate personally in practical learning situations.

We need, however, to recognise just as clearly that the extent to which we can expect them to be the masters of their own learning is limited, particularly in their early years in school. Learning should be seen as a partnership between child and teacher. This used not to be so: the dominant figure was the teacher. But in according to the child his rightful place in the partnership, the teacher must neither be excluded nor abdicate from hers. So it is perhaps worth summarising the responsibilities that her share of the partnership impose upon her.

1. She must give the children the security of a recognisable structure within which they may learn. They are entitled to this security, and the teacher is entitled to provide it.
2. Since mathematics is a sequential subject, the teacher has an obligation to present it systematically – whether in an integrated or a differentiated programme.
3. If the children are to be true partners in their learning, they must accept certain responsibilities and make some of their own decisions. But the question of *how much* responsibility and *which* decisions must be determined by the teacher and agreed by the child. This can happen only if the teacher trains the children towards these objectives, gradually extending the child's responsibility as he learns to accept it without anxiety.
4. The teacher is the professional, who knows what mathematical learning is suitable at a given level of the child's development. She must therefore make available the necessary learning opportunities, and structure them in a way that allows the child to benefit from them.

## THE TEACHER'S TIME

If child and teacher are to achieve success in their partnership, there are several aspects of organisation and management that may help and to which the teacher needs to give some thought. The first, and probably the most important of these, concerns making the most of the teacher's time. Since this is also the commodity that is likely to be in the shortest supply, the question is worth considering carefully. Throughout this book we have emphasised the importance of discussion, of *teaching* new mathematical processes to children and of following this up with teacher/child participation when the child puts the process into operation. These are all very time-intensive as far as the teacher is concerned, and there is a need to take a very close look at the problem.

We all know how easy it is to make it sound as though all you have to do is just organise everything efficiently, and you will double the time at your disposal without difficulty. But teaching is not like that. Neither, however, is it impossible for a teacher to organise her time so that a little more of it is available to do some of the things that help children to learn, by cutting down on some of the routine chores and interruptions that organisation and good training of the children may reduce. A little time saved here and there can make it just that bit more possible for a teacher to do more of the things that help to bring learning about.

Perhaps the first step is for her to look at her priorities and try to see if she can isolate a time pattern that will assist her. The actual pattern will be different according to whether she has a large class or a smaller one, whether the class is vertically grouped or there is a narrower age range, whether an integrated or a differentiated programme is in operation, and whether or not she has ancillary help for part of the time. However, though the time pattern will be different in all these circumstances, there will still be certain priorities that must be fitted into it.

We will therefore list the main duties that are time-intensive for the teacher during the school day. Because of the great variation in circumstances it cannot be an exhaustive list, and clearly it would be impractical to include the many incidentals like rubbing Jimmy's knee when he bangs it and intervening when Tony is about to cut off Anne's pigtail. The list includes only those aspects of her work in which forethought and deliberate planning may help the teacher to use her time more economically. These (which are not listed in a suggested order of importance) are:

1. Discussion with
   - (i) the class
   - (ii) a group
   - (iii) individual children
2. Giving direct teaching to
   - (i) the class

(ii) a group
(iii) individual children
3. Organising the activities of, and/or allocating tasks to
(i) the class
(ii) a group
(iii) individual children
4. Keeping the classroom in order; organising the distribution and clearing up of materials.
5. General 'supervisory' duties: milk distribution, changing for PE, keeping an eye on undirected activities.
6. Organising the work of the helper, if there is one.
7. Hearing reading.
8. Giving words for writing.
9. Encouraging, stimulating and helping with practical creative activities.
10. Assembling and removing displays; projects, children's work, wall charts.
11. Stimulating discussion among the children.
12. Marking.

Consider these twelve items. They are all time-intensive. They are all necessary. By no means all are directly concerned with teaching mathematics, but however important maths may be other things matter as well; and it is unrealistic to suppose that the infant teacher can allocate a disproportionate amount of her time to teaching maths.

One obvious truth emerges when we look at any list of the tasks which the infant teacher must undertake in the course of a day; and it is this. However strong the pressure on her to teach every child in a one-to-one relationship, she cannot, in practice, achieve this except on a very limited scale. Let us therefore review, realistically, some of the twelve items on the list.

How essential is it, in fact, or indeed educationally valuable, for discussion always to take place on a one-to-one basis? Sometimes it is, and nothing else will do. But if four or six children are all engaged in an activity that is based on the same mathematical principle, is there not real merit in talking with them as a group, instead of trying to do it at six different times of the day with six different children? Apart from saving the teacher's time, does group discussion not have the advantage of helping to stimulate an exchange of view among the children themselves, thereby nourishing further learning?

Discussion with the whole class will less often be as productive, but occasionally it is. It is so frequently implied that the teacher is guilty of some unforgivable sin if she does not discuss everything individually, or if she teaches in a group consisting of more than one child at a time. As generalisations, these implications are nonsense. Teachers should feel free to engage in group discussion and group teaching, both for their educational advantages and for their economy of time. This practice does not preclude individual or class discussion and teaching, when either is feasible and would be beneficial. However, one sometimes feels that there is a danger of isolating the child in making individual learning a ritual, instead of an objective which may best be attained within the social and intellectual structure of a community of children.

Items 3 and 4 on the list can be considerably assisted by secure organisational arrangements which help children to know what there is to do, and to make it possible for them to use some initiative (within the limits of their capacity) in helping themselves. This is why training, and a gradual extension of the children's responsibility, are not only educationally desirable but crucial if the teacher's time is to be used in more profitable ways than handing out assignments and looking for paint brushes. Organisation can be supportive without turning children into efficient automatons. (Not that there is much danger of this; children are not made that way, and anyone who thought in those terms would be defeated before they began!) However, there is no doubt that the teacher who has to spend her time on routine tasks which the children could well undertake for themselves will not have time to engage in much discussion or teaching. It really is a pretty straight choice between competent class management and teaching that is less effective than it would otherwise be.

The individual teacher should go through the rest of her list and determine her priorities in terms of how she may use her time to the children's best advantage.[1] No-one will pretend that such an exercise miraculously uncovers a great reservoir of unused time; and the larger the class, the more pressing is the problem. But the point we are making is that the teacher's only chance of finding some of the time she needs to talk with children, and to *teach* in the modern sense of the word, is to approach the problem positively and search out ways in which she can spend less time on this so that there is a little more to spend on that. The children need her to do this, if they are to learn as well as she would like them to.

## LEARNING MATHS IN AN INTEGRATED PROGRAMME

There are many interpretations of the word 'integrated', but we use it here to define a daily programme that is, to a greater or lesser degree, undifferentiated. That is to say, not every child would be doing maths at the same time. In a partially integrated programme certain parts of the day may be allocated to 3R work, or the arrangement may be taken much further and a child might do his maths at almost any time during the school day.

The greater the degree of integration of the programme, the more necessary it is for the teacher to take deliberate steps to ensure that essential learning experiences are not diluted or missed out altogether. The sequential nature of mathematics must still be respected, and it must still be systematically presented in progressive stages. Opportunity must be made for direct teaching, in groups and when necessary individually; and for the follow-up operations and the necessary discussion to take place, so that there can be a genuine intellectual interchange.

To achieve this, some teacher-direction *must* take place, and there is nothing inherent in modern progressive methods that denies the teacher the right to decision-making of this kind. Without question she should feel free to say: 'After play, I want Matthew, Mark, Luke and John to sit over here, because I'm going to teach you something new.' Having introduced the new process, she may then direct these four children to certain specific activities and experiences; and her system of records, which may include those that she trains the children to keep for themselves, must enable her to check that the new process is in fact followed up. Once this follow-up work has really got off the ground, each of the four children will continue at his own pace.

In practice, the teacher will find that Matthew and Mark will quite often decide to work together anyway, simply by virtue of the fact that they are engaged in something new which draws them together. Luke and John, perhaps, are less enthusiastic, or it may be that they currently have a stronger interest in something else. While respecting the other interest, the teacher must not allow the new bit of maths learning to be entirely neglected or the initial understanding will not be reinforced. So she makes sure that Luke and John do what is required, by directing them if necessary. They may also be encouraged to join up again with Matthew and Mark for a while. One way or another, the teacher must know what these children are doing, and must take the necessary steps to see that learning goes ahead and does not become lost in a welter of other occupations to which the child may be inclined to give too much priority.

Whatever the advantages of integration, those children who for one reason or another are not able to accept much responsibility for organising their time or making their own decisions must have clear teacher-direction for much of the day. Neither their sense of security nor their progress can be sacrificed to an ideology, and the intrinsic value of integrating the daily programme would be denied if this happened. But even with the benefits of a considerable degree of integration, and children who can profit by it, mathematics must still be taught so that it can be learned progressively and with understanding. The point is very well summarised in a report on primary mathematics:

'. . . the integrated day if not carefully planned and sensitively supervised, may leave large gaps in knowledge, may leave essential concepts unformed, and may fail to achieve any sort of progression. Where facility depends on practice, it may well hinder progress through failure to provide it. On the other hand, the integrated day could provide just that free atmosphere and stimulating situation in which children best thrive.'[2]

## PRACTICE

The question of giving the child practice in mathematical processes is a somewhat vexed one. It tends to conjure up images of the worst of the 'old maths', when children spent a good deal of time working through large numbers of sum cards of every kind; adding and subtracting, multiplying and dividing, not only numbers but feet and inches, ounces and pounds, shillings and pence and all the rest. All this took place in the happy belief that 'practice makes perfect' – not only perfect production of the right answers, but perfect understanding of the mathematics as well.

Times have changed since the bad old days, and sometimes the pendulum has swung so far to the opposite extreme that practice in any overt form has been roundly condemned as arid, unproductive and alien to the philosophy of modern educational methods. Most teachers have pitched their tents somewhere in between these opposing frontiers, but many experience a degree of uncertainty as to how far away they should move from the anti-practice or pro-practice territory.

It may help to resolve this dilemma if we first establish what we mean when we use the word 'practice'. Do we mean only that kind of activity in which pencil is committed to paper, for the purpose of recording mathematical processes sufficiently often to memorise them or to provide written evidence of accuracy? Or do we include in our definition practice of an experiential nature, which we believe may help to bring about understanding?

It is doubtful whether anyone would deny the value of experiential practice, of giving children many and varied experiences which are directed towards the abstraction of mathematical and other principles. This much we can accept without any nagging doubts. It is, however, the other kind of practice – the written kind – that is so often called in question.

The teacher's best guide here is her common sense. If an adult reads a book, the content of which he must particularly remember, he may well find it an advantage to make some written notes as he goes along. The very act of writing the notes helps to fix the content in his memory. We suggest that there is a parallel situation with children's learning. If a child were to have nothing but experiential practice, and were never required to record the results of his experience, he would be denied the benefit of 'making notes' to help him to identify what he has learned.

We have further guidance on this point from the researches of Dienes, whose fourth stage lays great emphasis on the need for children to record the results of their practical experience. According to Dienes, the child needs to represent the abstraction he has made in order to fix it and to make use of it in other situations. The representation does not always have to be a written one, but it would be difficult to argue that written representation should not take its place along with any other kind.

The question then arises of *how much* written representation – or practice in the written recording of the abstraction – is justifiable in these terms; and it is on this point that the teacher's common sense is her best ally. Too much written practice can be counter-productive. 'To give children more practice than they require to maintain efficiency is a waste of time.'[3] But there are different kinds of practice.

'Undoubtedly, practice is necessary, but there is a significant difference between practice that is mere repetition, and practice that reinforces a conceptual experience.'[4]

It all amounts to the fact that if practice is seen as a drill to *bring about* understanding, it has little educational value; but if it is seen as a means of reinforcing conceptual experience, it plays a constructive part in advancing children's learning. We should therefore not consign all sum cards and similar material to the dustbin. We should, however, make certain that they are used to consolidate conceptual understanding that has first been established by the other kind of practice – the experiential kind.

Both kinds of practice are needed. If the teacher understands the issues, and exercises a sense of moderation in her demands for the reinforcement of conceptual understanding, she will not go far wrong. Sum cards of various kinds are not 'out' for purposes of reinforcement; but neither are they 'in' as a means of promoting conceptual understanding in the first place.

## THE MATERIALS IN THE CLASSROOM

In the course of these chapters reference has been made to a variety of materials that help children to learn mathematics – materials such as Unifix cubes, Cuisenaire apparatus, Dienes Logiblocs, everyday items for sorting, weighing, and so on. It would be unnecessary to list them all over again, but with so much equipment in daily use the teacher will be conscious of the need for good classroom organisation in order that the right thing shall be available to the child at the right time and without confusion. This rests upon two considerations:

1. The physical layout of the classroom.
2. The storage and accessibility of the materials.

In both these respects the teacher's task is incomparably easier if space is not a pressing problem. If the physical layout of the classroom can include working bays, one of which is a maths area, there is much less difficulty in arranging for everything to be stored accessibly and in a way that is clear to the children. It is also easier for the children to help in keeping things in order. Certainly materials become disordered with use; it would be unnatural if they did not. But the children can straighten it all out now and again, and a reasonable degree of order can be maintained without anyone becoming neurotic about it.

However, the situation is very different when space is really tight; and there is no use pretending that this does not have adverse effects, not only on classroom organisation, but on learning itself. For example, there are clear mathematical advantages in having a shop, which are denied to children whose classroom is too small to accommodate one. If most materials have to be kept in boxes piled on top of each other, it is more difficult for a child to get at them and to replace them when an activity is finished. The most the teacher can do is to use what space there is as advantageously as possible, to label boxes clearly and identify them in a way that the non-reader can recognise and to help the children to keep it all in as orderly a way as can reasonably be expected of them.

But within the constraints of limitations which the teacher cannot alter, she should give careful attention to layout, storage and accessibility. The nearer she can get to making classroom organisation easy for the children, the less of a problem it will be to her, and learning experiences can be used to greater advantage. This is an obvious truth, but it needs stating out of sympathy for the teacher who is doing her best in difficult conditions, and who often feels that so much advice on mathematics teaching takes no account of all that she cannot do however much she would like to.

### CARD APPARATUS

Fairly detailed attention has already been given to the use of card apparatus, but the question of grading it remains to be considered. Once children reach the stage of being able to help themselves to this kind of material, they need to know which cards they should use and which they should not. This means that the grading system must be recognisable to the child as well as to the teacher.

There are many different ways of doing this, and the teacher should act in accordance with her own preference. As an example, however, we suggest a system of colour coding, because it is flexible, clear, and not very time-consuming to put into operation.

It is based on the use of self-adhesive coloured dots, packs of which are generally available. The teacher first decides upon the broad stages of development of the children in her class; these might, for example, be stages that apply to:

1. Children who are still in need of a good deal of pre-number experience.
2. Children who are moving towards an understanding of the conservation of number.
3. Those who have this understanding and are beginning to build upon it.
4. Those who are further ahead and are undertaking more complex operations.

Each of these stages is identified by a colour – say blue, white, red and yellow; and the cards which are appropriate to each stage have the coloured dots for that stage stuck on the back.

However, not all material within each broad stage will be at the same level. It will be at progressive levels, from that which is suitable for the child at the beginning of the stage to the kind he needs when he is very nearly ready to move from that stage to the next. One blue dot is therefore used for the first level within the stage, two blue dots for the next level and so on. It is a simple matter for the teacher to add more cards at each level as she sees the necessity for them, because problems such as upsetting a sequential numbering system do not arise.

There is no reason why the teacher should not also number some of the cards within the colour system if she wants to. This helps to avoid unnecessary repetition. With certain kinds of cards there is little value in the child doing the same one over and over again. Re-numbering may occasionally be necessary if more cards are added, but the problem is minimal because it does not often arise.

The disadvantage of a detailed grading system such as this is that if it is not administered in accordance with the needs of individual children it can become unnecessarily inflexible. Not every child should be required to do every card that is available for his level of development. One child may well need every card, and perhaps some additional material provided especially for him; another will move on much more quickly, and it will be evident that there is nothing to be gained from making him do the rest of the two dot cards before he can get on to those with three. Inflexibility is not a necessary evil of a clear grading system; but any system can be restricting if it is restrictively applied.

In conclusion, it is perhaps worth suggesting that teaching mathematics to young children requires as much understanding of the children as it does of the maths. It requires, also, a recognition that the child cannot learn it by discovering it all for himself, whether as an individual or as a member of a group. He will learn it, progressively and with understanding, only by being taught. It is upon his teacher's professional quality that he depends if his learning is to give him pleasure as well as intellectual enrichment. There need no longer be a dichotomy between children learning mathematics and children enjoying it.

## NOTES

1 For some suggestions about helping the time problem in hearing reading and giving words for writing, see Joy Taylor, *Reading and Writing in the First School* (Allen & Unwin, 1973), pp. 108–11 and 148–51.
2 K. L. Gardner, J. A. Glenn and A. I. G. Renton (eds), *Children Using Mathematics*: A Report of the Mathematics Section of the Association of Teachers in Colleges and Departments of Education (London: OUP, 1973), p. 119.
3 *The Story So Far* (Chambers and Murray for the Nuffield Foundation, 1969), p. 2.
4 *I Do and I Understand*, Nuffield Mathematics Teaching Project (Chambers and Murray, 1965), p. 8.

# Bibliography


Beard, R. M., *An Outline of Piaget's Developmental Psychology* (London: Routledge & Kegan Paul, 1969).

Biggs, Edith, *Mathematics for Younger Children* (London: Macmillan, 1971).

Biggs, J. B., *Mathematics and the Conditions of Learning* (Slough: National Foundation for Educational Research, 1967).

Brearley, Molly and Hitchfield, Elizabeth, *A Teacher's Guide to Reading Piaget* (London: Routledge & Kegan Paul, 1966).

Bruner, Jerome S., *Beyond the Information Given* (USA: 1973; London: George Allen & Unwin, 1974).

Bryant, P. E. and Trabasso, T., 'Transitive Inferences and Memory in Young Children', *Nature*, vol. 232 (August, 1971).

Dienes, Zoltan P., *The Six Stages in the Process of Learning Mathematics* (Paris: OCDL, 1970). Published in English translation (trans. P. L. Seaborne), NFER, 1973.

Dienes, Z. P. and Golding, E. W., *Learning Logic, Logical Games* (4th impression) (Harlow: The Educational Supply Association, 1970).

Fletcher, Harold (ed.), Teacher's Resource Book, *Mathematics for Schools*, Level I (London: Addison-Wesley Publishers Ltd, 1970).

Frobisher, Beryl and Gloyn, Susan, *Infants Learn Mathematics* (London: Ward Lock Educational, 1969).

Gardner, K. L., Glenn, J. A. and Renton, A. I. G., *Children Using Mathematics* (Oxford University Press for The Mathematics Section of the ATCDE, 1973).

Holt, Michael and Dienes, Zoltan, *Let's Play Maths* (Harmondsworth: Penguin Books Ltd, 1973).

Isaacs, Nathan, *The Growth of Understanding in the Young Child* (3rd impression) (London: Ward Lock Educational, 1964).

Isaacs, Nathan, *New Light on Children's Ideas of Number* (6th impression) (London: Ward Lock Educational, 1966).

Lovell, K., *The Growth of Basic Mathematical and Scientific Concepts in Children* (3rd impression) (London: University of London Press, 1971).

Nuffield Mathematics Project
*Mathematics Begins* (1967)
*Beginnings* (1968)
*Pictorial Representation* (1970)
*I Do and I Understand* (1965)
*The Story So Far* (1969)
*Mathematics: the first 3 years* (2nd impression, 1973) (London: Chambers and Edinburgh, Murray, for the Nuffield Foundation).

Plowden Report, *Children and Their Primary Schools* (London: HMSO, 1967).

*Primary Mathematics: a further report* (Mathematical Association, 1970).
Schools Council, *Metres, Litres and Grams* (London: Evans/Methuen Educational, 1971).
Schools Council Curriculum Bulletin No. 1, *Mathematics in Primary Schools* (London: HMSO, 1972).
The Royal Society, *Metric Units in Primary Schools* (London: 1969).
Thyer, Dennis and Maggs, John, *Teaching Mathematics to Young Children* (London: Holt, Rinehart & Winston, 1971).
Williams, E. M. and Shuard, Hilary, *Primary Mathematics Today* (London: Longman, 1970).
Williams, J. D., *Teaching Technique in Primary Maths* (Slough: National Foundation for Educational Research, 1971).

# Index

Coláiste Oideachais Mhuire Gan Smál
Luimneach